ENVIRONMENTAL HEALTH

Indoor Exposures, Assessments and Interventions

ENVIRONMENTAL HEALTH

Indoor Exposures, Assessments and Interventions

Edited by
Theodore A. Myatt, ScD, and Joseph G. Allen, DSc

Apple Academic Press
TORONTO NEW JERSEY

Apple Academic Press Inc.
3333 Mistwell Crescent
Oakville, ON L6L 0A2
Canada

Apple Academic Press Inc.
1613 Beaver Dam Road, Suite # 104
Point Pleasant, NJ 08742
USA

Exclusive worldwide distribution by CRC Press, a Taylor & Francis Group

International Standard Book Number: 978-1-926895-20-8 (Hardback)

Printed in the United States of America on acid-free paper

Library of Congress Control Number: 2012951936

Library and Archives Canada Cataloguing in Publication

Environmental health : indoor exposures, assessments and interventions/edited by Theodore A. Myatt and Joseph G. Allen.

Includes bibliographical references and index.
ISBN 978-1-926895-20-8
1. Indoor air quality--Case studies. 2. Environmental health--Case studies. I. Myatt, Theodore A. II. Allen, Joseph G., 1975-

TD883.17.E58 2013 613.5 C2012-906389-4

About the Editors

Theodore A. Myatt, ScD

Dr. Ted Myatt is the Director of the Partners Healthcare Institutional Biosafety Committee and Research Compliance Manager at Brigham and Women's Hospital in Boston, Massachusetts. Dr. Myatt received his doctorate of science in environmental health from the Harvard School of Public Health, where is research focused on evaluating exposures to respiratory viruses. He also received a master's degree in environmental management from the Nicholas School of Environment at Duke University. Additionally, Dr. Myatt is an instructor of environmental science at Brandeis University and is a senior scientist with Environmental Health and Engineering, Inc., where he focuses on evaluating exposures to biohazardous agents.

Joseph G. Allen, DSc

Dr. Joseph G. Allen is a senior scientist with Environmental Health and Engineering, Inc., in Needham, Massachusetts, and Group Manager for the Advanced Analytics and Building Science Division. He received his doctorate of science and masters of public health degrees from Boston University. Dr. Allen concentrates his research on exposure science, statistical analysis of environmental and occupational data, and environmental epidemiology. He has directed community and occupational exposure assessments related to the broad class of environmental health topics, including persistent organic compounds, flame retardants, VOCs, heavy metals, air pollution, and infectious disease outbreaks.

Contents

List of Contributors

Robin M. Ackerman
Simmons College, Department of Nursing, Boston, USA

Joseph G. Allen
Environmental Health & Engineering, Inc., Needham, Massachusetts, USA

Larisa Altshul
Department of Environmental Health, Exposure, Epidemiology and Risk Program, Harvard School of Public Health, 665 Huntington Avenue, Boston, Massachusetts, USA

Udeni K. Alwis
Harvard School of Public Health, Boston, Massachusetts, USA

Peter J. Ashley
Department of Housing and Urban Development, Office of Healthy Homes and Lead Hazard Control, 451 Seventh Street, SW, Room 8236, Washington, DC, USA

Anilo Bello
Department of Work Environment, University of Massachusetts, Lowell, USA

Asa Bradman
Center for Environmental Research and Children's Health (CERCH), School of Public Health, University of California, 1995 University Avenue Suite 265, Berkeley, California, USA

Doug Brugge
Public Health and Community Medicine, Tufts University, 136 Harrison Avenue, Boston, Massachusetts, USA

Antonia M. Calafat
Division of Laboratory Sciences, National Center for Environmental Health, Centers for Disease Control and Prevention, Atlanta, Georgia, USA

Sherry L. Dixon
The National Center for Healthy Housing, Columbia, Marylan, USA

John L. Durant
Department of Civil and Environmental Engineering, Tufts University, Medford, Massachusetts, USA

Susan M. Duty
King's College London, Wolfson Centre for Age Related Diseases, Guy's Campus, London, UK

Brenda Eskanazi
Center for Environmental Research and Children's Health (CERCH), School of Public Health, University of California, 1995 University Avenue Suite 265, Berkeley, California, USA

M. Patricia Fabian
Harvard School of Public Health, Department of Environmental Health, Boston, Massachusetts, USA

Jeannette Ferber
Center for Environmental Research and Children's Health (CERCH), School of Public Health, University of California, 1995 University Avenue Suite 265, Berkeley, California, USA

Joanna M. Gaitens
University of Maryland School of Medicine, 405 West Redwood Street, Baltimore, Maryland, USA

Diane R. Gold
Channing Laboratory, Department of Medicine, Brigham and Women's Hospital, Boston, Massachusetts, USA; and Harvard Medical School, Boston, Massachusetts, USA

Martha E. Harnly
California Department of Public Health, Environmental Health Investigations Branch, 850 Marina Bay Parkway P-3, Richmond, California, USA

Robert F. Herrick
Department of Environmental Health, Harvard School of Public Health, 665 Huntington Ave., Boston, Massachusetts, USA

Elaine Hoffman
Harvard School of Public Health, Boston, Massachusetts, USA
Allan Hubbard
Division of Biostatistics, School of Public Health, University of California, Berkeley 50 University Hall, MC 7356, Berkeley, California, USA

David E. Jacobs
US Department of Housing and Urban Development, Office of Healthy Homes and Lead Hazard Control, Washington, D.C., USA

Matthew H. Kaufman
Environmental Health & Engineering, Inc., Needham, Massachusetts, USA

David L. MacIntosh
Environmental Health & Engineering, Inc., Needham, Massachusetts, USA

James J. McDevitt
Harvard School of Public Health, Department of Environmental Health, Boston, Massachusetts, USA

Thomas E. McKone
Center for Environmental Research and Children's Health (CERCH), School of Public Health, University of California, 1995 University Avenue Suite 265, Berkeley, California, USA; and Lawrence Berkeley National Laboratory, One Cyclotron Road, Berkeley, California, USA

John D. Meeker
Department of Environmental Health Sciences, University of Michigan School of Public Health, 109 S. Observatory St, Ann Arbor, Michigan, USA

Donald K. Milton
Channing Laboratory, Department of Medicine, Brigham and Women's Hospital, Boston, Massachusetts, USA; and University of Maryland, College Park, Maryland, USA

Taeko Minegishi
Environmental Health & Engineering, Inc., Needham, Massachusetts, USA

Theodore A Myatt
Environmental Health & Engineering, Inc., Needham, Massachusetts, USA

Jyothi Nagaraja
Battelle Memorial Institute, 505 King Avenue, Columbus, Ohio, USA

Marcia Nishioka
Battelle Memorial Institute, 505 King Avenue, Columbus, Ohio, USA

Melissa J. Perry
Occupational Health Program, Harvard School of Public Health, Boston, Massachusetts, USA

Margaret M. Quinn
Department of Work Environment and the Lowell Center for Sustainable Production, University of Massachusetts Lowell, Lowell, Massachusetts, USA

Lesliam Quiros-Alcalá
Center for Environmental Research and Children's Health (CERCH), School of Public Health, University of California, 1995 University Avenue Suite 265, Berkeley, California, USA

Christine Rioux
Department of Public Health and Community Medicine, Tufts University, Boston, Massachusetts, USA

Joanne E. Sordillo
Channing Laboratory, Department of Medicine, Brigham and Women's Hospital, Boston, Massachusetts, USA; and Harvard Medical School, Boston, Massachusetts, USA

Heather M. Stapleton
National Institute of Standards and Technology, Gaithersburg, Maryland, USA

Warren Strauss
Battelle Memorial Institute, 505 King Avenue, Columbus, Ohio, USA

Jonathan W. Wilson
National Center for Healthy Housing, Columbia, Maryland, USA

List of Abbreviations

2–BE	2–Butoxyethanol
ACH	Air changes per hours
ACS	American Cancer Society
AER	Air exchange rates
AH	Absolute humidity;
AHU	Air-handling units
BC	Black carbon
CADR	Clean air delivery rate
CDPR	California Department of Pesticide Regulation
CO	Carbon monoxide
cPAD	Chronic adjusted population dose
DF	Detection frequency
DRI	Direct reading instrument
EC	Elemental carbon
EPA	Environmental Protection Agency
ETS	Environmental tobacco smoke
FEV1	Forced expiratory volume in 1 second
FEV1/FVC	Ratio of FEV1 and forced vital capacity
FEV25–75	Forced expiratory volume between 25 and 75
FVC	Forced vital capacity
GEE	Generalized estimating equation
HEPA	High efficiency particle arrestance
HQ	Hazard quotient
HVS3	High volume small surface sampler
Koc	Water:organic carbon partition coefficient
Kow	Octanol:water partition coefficient
LOD	Limit of detection
m	Meters
mb	Millibar
MSDSs	Materials safety data sheets
ng/g	Nanogram per gram
NHANES	National Health and Nutrition Examination Survey
NIOSH	National Institute for Occupational Health and Safety
NIST	National Institute of Standards and Technology
NO_2	Nitrogen dioxide
NOx	Oxides of nitrogen
OP	Organophosphorous
OR	Odds ratio
OSHA	Occupational Health and Safety Administration
PBO	Piperonyl butoxide
PCA	Principal components analysis

PCB	Polychlorinated biphenyls
PDI	Potential daily intake
PEL	Permissible exposure levels
pg/g	Picogram per gram
PID	Photo ionization detector
PM	Particulate matter
$PM_{2.5}$	Particles less than 2.5 microns
PM10	Particulate matter less than 10 um
PPAH	Particle bound polyaromatic hydrocarbons
PON1	Paraoxonase 1
PUR	Pesticide use reporting
qPCR	Quantitative polymerase chain reaction
REL	Recommended exposure limit
RfD	Reference dose
RH	Relative humidity
RT-qPCR	Reverse transcription-quantitative polymerase chain reaction
SARS	Severe acute respiratory syndrome
SES	Socioeconomic status
SO_2	Sulfur dioxide
SRS	Surrogate recovery standard
TLV	Threshold limit values
TVOC	Total volatile organic compounds
TWA	Time weighted average
UFP	Ultra fine particles
ug/m^3	Micrograms per cubic meter of air
um	Micrometers
veh/d	Vehicles per day
veh/h	Vehicles per hour
VOC	Volatile organic compounds

Introduction

People spend 87% of their time indoors, yet we often think of our exposure to pollutants of outdoor origin first—such as air pollution—and neglect to consider what is happening while we are inside our homes, schools, or offices, seemingly protected from outdoor elements. Research focusing on the indoor environments demonstrates that building occupants are exposed to a mixture of pollutants of chemical and biological origin. The sources of these pollutants are varied; they may be emitted from building materials or consumer products, introduced via outdoor air infiltration, or shed from building occupants. For pollutants with indoor sources, numerous studies have shown that airborne concentrations are much greater indoors. For pollutants that enter buildings from outdoors, such as particulate matter from combustion sources or fungal spores, the indoor environment contributes more to a person's total exposure due to people spending more time indoors than outdoors. Regardless of the source, the exposure pathways or the levels observed for these pollutants are in many cases unique to the indoor environment.

The sources for the pollutants found indoors are as varied as the pollutants themselves. In some cases, the pollutant is intentionally added to an indoor material, such as lead to paint or flame-retardants to furniture and electronics. In other cases, the pollutant is ubiquitous, but indoor conditions allow it to flourish, such as indoor mold. Pollutants may be introduced either directly by the occupants, as is the case with infectious diseases that are expelled from occupants, like influenza virus, or indirectly, through decisions of the occupants, such as the choice to smoke in the home or to have an animal that sheds allergens.

The built environment can impact the exposure to many of the pollutants found indoors. For example, buildings with tight envelopes that have low ventilation rates may minimize pollutants of outdoor origin from entering the building, but allow pollutants with indoor sources to build up to high concentrations. The age of the building may impact both the ventilation rate of the building and the type and quantity of building materials and consequently, the pollutants released from these materials.

In this collection of studies, we focus on exposure to pollutants that occur indoors, typically, but not exclusively, in residences. Recent advances have been made in identifying the pollutants in indoor environments, the health effects associated with indoor or personal exposures, and interventions that can be implemented by typical occupants to mitigate exposures. A variety of methods can be employed to assess building occupant exposures to indoor pollutants. Studies may directly measure the indoor air, as can be done for a variety of particulate and gaseous pollutants, to assess the inhalation route of exposure. Alternatively, pollutants may be evaluated in dust samples to characterize exposure via dust ingestion, which can occur directly (e.g., a toddler mouthing a toy), through incidental ingestion (i.e., accumulated dust on hands is transferred into the body while eating "finger foods") or, for some compounds, through dermal absorption.

Dust sampling has long been used as a method to assess exposures to lead. While lead has been banned in paint and gasoline for decades, it remains a legacy pollutant because it is in the paint in older homes; hundreds of thousands of children in the United States have blood lead levels that exceed the level where learning and behavioral effects have been well documented to occur. Gaitens and colleagues evaluated lead levels in dust samples collected in hundreds of homes as part of the National Health and Nutrition Examination Survey (NHANES) in "Lead Exposure of U.S. Children from Residential Dust." This study shows that home characteristics and demographic information are strong predictors of dust lead levels.

Another legacy pollutant is polychlorinated biphenyls (PCBs), which were banned in the late 1970s but have reemerged as a current issue. Research on (PCBs) has established associations between PCB exposure and effects on the immune, reproductive, nervous, and endocrine systems. PCBs can be found in a variety of products used in homes, such as adhesives, caulking materials, and heat insulation, and are now commonly being found in schools built before PCBs were banned. In "Teachers Working in PCB-Contaminated Schools," Herrick and colleagues report that older teachers had significantly higher total PCB concentrations in their blood serum compared to the general population of a comparable age, again demonstrating the impact of the indoor environment on exposure.

While lead is a pollutant that has been known to cause health impacts for centuries, and PCBs were banned decades ago due to health concerns, there is a considerable amount of research focusing on indoor exposures and potential health effects of newer classes of pollutants, such as flame-retardants. Meeker and Stapleton report in their chapter, "Flame-Retardants' Effect on Hormone Levels and Semen Quality" that newer classes of flame-retardants, such as organophorous flame-retardants, can be identified in nearly every home investigated. While much research remains to be done, this report indicates that exposure to organophorous flame-retardants in the dust of our homes is ubiquitous and may be related to endocrine and reproductive effects.

Just as Gaitens and colleagues report that home characteristics are predictors of dust lead levels, Sordillo and colleagues demonstrate that home characteristics can explain some of the variability of bacterial and fungal levels. In "Bacterial and Fungal Microbial Biomarkers in House Dust," the authors evaluated whether home characteristics could be used as a proxy for expensive measurements of bacterial and fungal levels. Sordillo and colleagues report that home characteristics related to dampness were significant predictors of microbial exposure.

An important aspect of a building is its location to outdoor pollutant sources. As Brugge, Durant and Rioux note in their chapter titled "Pollutants from Vehicle Exhaust Near Highways," approximately 11% of homes in the U.S. are within 100 meters of a highway. Their chapter reviews the substantial literature on the health effects of those that live near highways and shows that these primarily indoor exposures to traffic-related pollutants are associated with development of asthma and reduced lung function in children and cardiac and pulmonary mortality.

As mentioned above, exposure to environmental agents commonly found in the indoors can lead to the development of asthma and asthma exacerbations in sensitive populations. In many cases, these agents are ubiquitous and exposures cannot be completely prevented. Methods to minimize or control exposures, especially in the residential environment, could have profound impacts on asthma symptom reduction and possibly even health-care costs. In research described in "Asthma Triggers in Indoor Air," the use of enhanced filtration in homes substantially reduced levels of common asthma triggers, such as cat allergen, fungal spores, and environmental tobacco smoke.

Asthma can also be triggered by respiratory viruses such as rhinovirus and influenza virus. Transmission of respiratory viruses is thought to occur most commonly in indoor environments due to the close proximity of occupants, and potentially elevated concentrations in indoor air. The chapter "Home Humidification and Influenza Virus Survival" shows that modification to the indoor environment, through the use of humidifier to raise the humidity in the indoor air, may reduce the survival of influenza virus in the air and on surfaces, thus limiting exposures to occupants.

Consumer products can also be an overlooked source of exposure to chemical pollutants that have potentially important health implications. In "Airborne Exposure from Common Cleaning Tasks," Anila Bello and colleagues advance the field by conducting a quantitative exposure assessment of total volatile organic compounds (TVOCs) and specific volatile compounds. Their research demonstrates that airborne exposures during short-term cleaning tasks can approach occupational exposure limits.

Another example of consumer products influencing indoor exposure is the widespread use of plasticizers, many of which are found in consumer products used in the home, and some have been associated with health impacts through endocrine disruption. Phthalates are chemicals used as plasticizers, solvents, and stabilizers for a wide variety of products, including personal-care products such as hair spray, body lotion, fragrances, and deodorant. In the chapter "Phthalate Monoesters from Personal Care Products," Duty and colleagues demonstrated that the uses of these products are important contributors to the body burden of phthalates.

Occupants also introduce into homes products, such as pesticides, that are well understood to have negative health impacts if elevated exposures occur. In "Pesticides in House Dust," Quiros-Alcala and colleagues measured dust in low-income homes for multiple insecticides. More than one-half of the study participants reported use of insecticides indoors. While the authors conclude that the exposures to children in the homes do not exceed health-based standards, the children may still be at risk of health effects due to these indoor exposures.

In this collection of research, we focused on studies that highlight the significance of the indoor environment to exposure and health, for both old and new pollutants, and those of chemical and biological origin. From research on well-known pollutants like lead, mold, and PCBs, to the newly identified exposures of phthalates and flame-retardants, exposures that occur indoors are increasingly being shown to be highly relevant in terms of total exposure to pollutants—and in many cases, these exposures can result in adverse health impacts.

— Theodore A. Myatt, ScD
Joseph G. Allen, DSc

1 Lead Exposure of U.S. Children from Residential Dust

*Joanna M. Gaitens, Sherry L. Dixon,
David E. Jacobs, Jyothi Nagaraja, Warren Strauss,
Jonathan W. Wilson, Peter J. Ashley*

CONTENTS

INTRODUCTION

Lead-contaminated house dust is a major source of lead exposure for children in the United States. In 1999–2004, the National Health and Nutrition Examination Survey (NHANES) collected dust lead (PbD) loading samples from the homes of children 12-60 months of age. In this study we aimed to compare national PbD levels with existing health-based standards and to identify housing and demographic factors associated with floor and windowsill PbD. We used NHANES PbD data (n = 2,065 from floors and n = 1,618 from windowsills) and covariates to construct linear and logistic regression models.

The population-weighted geometric mean floor and windowsill PbD were 0.5 μg/ft^2 [geometric standard error (GSE) = 1.0] and 7.6 μg/ft^2 (GSE = 1.0), respectively.

Only 0.16% of the floors and 4.0% of the sills had PbD at or above current federal standards of 40 and 250 μg/ft^2, respectively. Income, race/ethnicity, floor surface/condition, windowsill PbD, year of construction, recent renovation, smoking, and survey year were significant predictors of floor PbD [the proportion of variability in the dependent variable accounted for by the model (R^2) = 35%]. A similar set of predictors plus the presence of large areas of exterior deteriorated paint in pre-1950 homes and the presence of interior deteriorated paint explained 20% of the variability in sill PbD.

Most houses with children have PbD levels that comply with federal standards but may put children at risk. Factors associated with PbD in our population-based models are primarily the same as factors identified in smaller at-risk cohorts. PbD on floors and windowsills should be kept as low as possible to protect children.

The U.S. Centers for Disease Control and Prevention (CDC) estimates that 310,000 children between 1 and 6 years of age in the United States have blood lead (PbB) levels > 10 micrograms per deciliter [1]. The health effects associated with PbB levels at or above this level of concern have been well documented, including learning and behavioral problems [2]. Evidence suggests that children with PbB < 10 μg/dL also experience notable adverse effects and that no safe level of lead exposure exists [3–6]. In this study we identified factors associated with childhood residential dust lead (PbD) exposure.

Lead exposure can occur through a variety of sources, including air, bare soil, home remedies, drinking water, toy jewelry, and others [7]. However, the major pathway of exposure for children is from deteriorated lead-based paint and lead-contaminated dust in the home that is ingested during normal hand-to-mouth behavior [1, 8]. The importance of PbD from lead paint was recognized very early [9], and work was done subsequently in an attempt to quantify its exposure contribution [10].

Although lead-based paint was banned from residential use in 1978, approximately 38 million older housing units in the United States still contained lead-based paint, and an estimated 24 million housing units contain significant lead hazards as of 2000 [11]. Although intact paint does not generally result in significant immediate exposure, all paint eventually deteriorates; lead-based paint that is chipping, peeling, or flaking or otherwise separating from its substrate presents a hazard. In addition, lead-contaminated settled dust, which is often found in houses with deteriorated lead-based paint, is a significant lead hazard [8]. PbD can also be generated from the friction and impact of lead-painted surfaces [12] and during housing renovation and repair projects where lead-based paint is present and proper precautions are not in place [5, 13]. The use of leaded gasoline, which peaked in the early 1970s, has also contaminated soil around the home [14]. Many studies, employing a variety of research designs, have demonstrated that soil-lead concentrations are a significant contributor to PbD and children's PbB [8, 12, 15, 16]. Numerous cross-sectional [8] and longitudinal studies [17] have firmly established the correlation of settled PbD and children's PbB. In an effort to protect young children from adverse effects of lead, current federal health-based hazard standards indicate that floor and window PbD should not exceed 40 μg/ft^2 and 250 μg/ft^2, respectively [18].

Through an interagency agreement with the CDC, the U.S. HUD Office of Healthy Homes and Lead Hazard Control sponsored the collection of PbD wipe samples and housing-related data through the National Health and Nutrition Examination Survey (NHANES) from 1999 through 2004, marking the first time that NHANES has collected both health and housing environmental data. Using these national survey data, we investigated PbD in homes to explore the feasibility of lowering PbD standards. Here we present the demographic and housing characteristics associated with floor and windowsill PbD. We used linear regression modeling to predict natural log-transformed floor and windowsill PbD and logistic regression modeling to predict the log odds of PbD at various levels. A companion article in this issue [12] presents the analysis of NHANES data with respect to childhood PbB levels. Together these data identify the important risk factors and the relationship between PbD and children's PbB in the United States in recent years.

METHODS

Study Population

We analyzed data from three waves of NHANES (1999–2000, 2001–2002, 2003–2004). NHANES is a nationally representative cross-sectional household survey that uses a complex, stratified, multistage probability sampling design to track the health of the noninstitutionalized civilian U.S. population. It has been a primary source of information about the national distribution of children's PbB. Details of the NHANES protocol and all testing procedures are available elsewhere [19–21]. Our data set included 2,155 children 12 to 60 months of age with measured PbB. Only children living in housing built before 1978, when the United States banned the use of lead in residential paint, were included in the analysis of the influence of floor PbD on children's PbB (n = 731).

Child, Household, and Housing Characteristics

NHANES interviewers collected data on age, race/ethnicity, sex, socioeconomic measures [family and household income and poverty-to-income ratio (PIR)], and other self-reported health data through a structured household interview. Participants self-reported their race and ethnicity. In this analysis, we used a composite race/ethnicity variable: non-Hispanic white, non-Hispanic black, Hispanic, or other race. These variables, as well as the housing characteristic variables, are described in the companion article. The PIR is the ratio of income to the family's poverty threshold [22]. PIR values < 1.00 are below the poverty threshold, whereas PIR values of ≥ 1.00 indicate income above the poverty level. Variables on smoking behavior included the presence of smoking in the home, number of smokers, and the number of cigarettes smoked in the home per day. During their visit to the mobile examination center, NHANES participants provided venous blood samples, which were analyzed for PbB, serum cotinine, ferritin, iron, and total iron binding capacity.

NHANES measured PbB using graphite furnace atomic absorption spectrophotometry. The laboratory detection limit (DL) was 0.3 μg/dL. Only 0.23% of the sample results were below the DL. The DLs for cotinine were 0.05 ng/mL and 0.015 ng/mL for 1999–2000 and 2003–2004, respectively. For 2001–2002, there was a mixture of these two DLs. Twenty-six percent of the cotinine samples were below the DL. For all NHANES laboratory measurements, results below the DL were assigned the value of

$$L/\sqrt{2}$$

Statistical Methods

Data were analyzed using SUDAAN (version 9.0.0; RTI International, Research Triangle Park, NC) and SAS (SAS System for Windows, version 9.1.3; SAS Institute Inc., Cary, NC). We used a linear regression model to predict natural log-transformed PbB and logistic regression models to predict the probability that a child's PbB exceeded either 5 or 10 µg/dL. The models adjusted the parameter estimates for the clustering and unequal survey weights within NHANES. The modeling employed Taylor series expansion theory without degrees of freedom adjustments. Backward elimination of insignificant independent variables ($p > 0.10$) was followed by additional steps to allow addition and/or removal of variables. To provide an accurate prediction of children's PbB without eliminating large fractions of the study sample because of missing values, we fit an intercept term for each variable that had a missing value. The overall p-value is the type 3 F-test that captures the overall statistical significance of each variable included in the model. For categorical variables with missing values, the missing level was not included in this hypothesis test.

Because NHANES collected serum cotinine only for children ≥ 3 years of age, many more children had questionnaire-based smoking data available than serum cotinine measurements. Therefore, we gave questionnaire-based smoking variables priority over measured serum cotinine levels.

Geometric mean (GM) PbB peaks between 18 and 36 months of age and slowly declines over the next few years, with the rate of decline varying in different populations [23–25]. Based on the relationships between age and PbB observed in these studies, we determined that a quartic function of age of the child fit best.

Although most other analyses of the relationship between log PbB and log floor PbD were based on a linear relationship, the relationship may not be linear across the relatively low ranges observed in NHANES [8, 26, 27]. To investigate this further, we analyzed other data sets: the National Evaluation of the HUD Lead-Based Paint Hazard Control Grant Program (the Evaluation) [28, 17]; the Rochester Lead-in-Dust Study (Rochester) [29, 30]; and the HUD National Risk Assessment Study (the RA Study) [31]. For each of these data sets and NHANES, we predicted log-PbB based on a cubic function of log floor PbD for children < 6 years of age (Table 1). The NHANES model accounted for clustering and unequal survey weights.

We predicted PbB at different PbD levels for children living in homes built before 1978 while controlling for other predictors of PbB using the aforementioned linear and logistic regression models and the population-weighted averages of covariates (except floor and sill PbD). For categorical variables, the levels were weighted according to their population-weighted relative frequency distribution. For continuous covariate variables with intercepts fit for missing values, the same percent of missing values observed in the population was assumed for the average risk values. For windowsill PbD values, we used a linear regression based on unweighted data from homes built before 1978 ($n = 601$). The correlation coefficient for the linear relationship between natural log-transformed sill and floor PbD is 0.38 ($p < 0.001$). The regression equation is: ln(sill PbD) = 2.654+0.524 × ln(floor PbD) ($r = 0.38$, mean square error = 2.78; SE for the intercept and slope are 0.070 and 0.053, respectively).

TABLE 1 Models Predicting Children's Log Pbb Based on Floor Pbd.

		Data set			
Statistic	**Term**	**Evaluations**	**NHANES**	**RA Studyb**	**Rochesterc**
Regression	Intercept	1.664 (0.073)	0.826 (0.023)	0.938 (0.193)	1.168 (0.194)
Coefficient (SE)	Log (floor PbD)	0.269 (0.042)	0.319 (0.029)	0.491 (0.293)	0.340 (0.103)
		$p < 0.001$	$p < 0.001$	$p = 0.096$	$p = 0.003$
	$[\text{Log (floor PbD)}]^2$	−0.022 (0.006)	0.033 (0.008)	0.003 (0.117)	−0.021 (0.012)
		$p = 0.001$	$p < 0.001$	$p = 0.980$	$p = 0.083$
	$[\text{Log (floor PbD)}]^3$	—	−0.014 (0.004)	−0.009 (0.013)	—
			$p < 0.001$	$p = 0.498$	
Overall *p*-value for log (floor PbD)		$p < 0.001$	$p < 0.001$	$p < 0.001$	$p < 0.001$
R^2		6.9	23.6	23.3	8.6
Mean-square error		0.512	0.262	0.532	0.350
No. of children/ units		1,096	2,065	203	205

[a]Data from Galke et al. (2001), U.S. HUD (2004).
[b]Data from Wilson et al. (2007).
[c]Data from Lanphear et al. (1996a, 1996b).

The GM PbB and the probability that PbB is ≥ 10 μg/dL and ≥ 5 μg/dL were predicted for floor PbD ranging from 0.25 to 40 μg/ft² using the linear and logistic regression models, respectively. Although exponentiation of the predicted logarithm of the PbB may slightly overestimate the expected GM PbB, the large sample size minimizes the overestimation [32].

RESULTS

Characteristics of the Study Population

PbB data were available for 2,155 children 12–60 months of age. The population-weighted GM PbB was 2.03 μg/dL. Eight percent were ≥ 5 μg/dL, 1.71% were ≥ 10 μg/dL, and 0.33% were ≥ 15 μg/dL. Gaitens et al. (2009) present the descriptive statistics for PbD and additional housing variables. Here we present descriptive statistics for variables found to be significant ($p < 0.10$) in the PbB model (Tables 2 and 3). The weighted distribution shows that approximately 57% of the sampled population was non-Hispanic white, 15% was non-Hispanic black, and 24% was Hispanic. The vast majority (97.43%) of the children were born in the United States. Fifty-eight percent lived in a single-family detached house, and almost one-quarter lived in an apartment. Fifty-two percent of the homes for which data on the year of construction were available were built before 1978. Approximately 6% of homes were built before 1950 and had evidence of deteriorated paint (i.e., peeling, flaking, or chipping paint) inside. Ten percent of children lived in pre-1978 homes where window, cabinet, or wall renovation was completed in the preceding 12 months.

TABLE 2 Descriptive Statistics for PbB, Housing, and Demographic Variables, NHANES 1999–2004.

		All homes			Pre-1978 homes		
			Weighted			Weighted	
Variable	Level	No.	GM (GSE)	AM (SE)	No.	GM (GSE)	AM (SE)
PbB (μg/dL)	—	2,155	2.03 (1.03)	2.51 (0.09)	731	2.16 (1.03)	2.69 (0.10)
Age (months)	—	2,155	33.6 (1.01)	36.7 (0.35)	731	33.4 (1.02)	36.6 (0.64)
Cotinine (ng/mL)	Missing	1,326	—	—	449	—	—
	Nonmissing	829	0.18 (1.14)	1.02 (0.11)	282	0.18 (1.18)	0.97 (0.20)
Floor surface/conditiona × floor PbD (μg/ft²)	Missing	90	—	—	0	—	—
	Not smooth and cleanable	25	1.70 (1.47)	4.92 (2.11)	8	1.26 (1.69)	4.67 (3.60)
	Smooth and cleanable or carpeted	2,040	0.52 (1.05)	1.34 (0.14)	723	0.64 (1.07)	1.78 (0.31)
	All nonmissing	2,065	0.52 (1.05)	1.34 (0.14)	731	0.64 (1.07)	1.80 (0.31)
PIRb	Missing	136	—	—	24	—	—
	Nonmissing	2,019	—	2.07 (0.05)	707	—	2.25 (0.09)
Windowsill PbD (μg/ft²)	Missing	537	—	—	130	—	—
	Nonmissing	1,618	7.64 (1.07)	57.8 (9.42)	601	10.5 (1.11)	71.8 (14.8)

Abbreviations: AM, arithmetic mean; GSE, geometric standard error.

[a]Table 1 in the companion article presents descriptive statistics by the expanded groups of floor surface/condition.

[b]GM and GSE are undefined because of zero values.

TABLE 3 Descriptive Statistics for PbB, Housing, and Demographic Categorical Variables, NHANES 1999–2004.

		All homes			Pre-1978 homes		
			Weighted percent			Weighted percent	
Variable	Levels	No.	Missing included	Missing excluded	No.	Missing included	Missing excluded
PbB ≥ 5 μg/dL	No	1,918	91.88	91.88	643	90.84	90.84
	Yes	237	8.12	8.12	88	9.16	9.16
PbB ≥ 10 μg/dL	No	2,104	98.29	98.29	708	97.97	97.97
	Yes	51	1.71	1.71	23	2.03	2.03
PbB ≥ 15 μg/dL	No	2,140	99.67	99.67	725	99.65	99.65
	Yes	15	0.33	0.33	6	0.35	0.35
Home-apartment type	Missing	39	1.77	—	7	0.47	—
	Mobile home or trailer	205	9.77	9.95	20	2.69	2.70
	One-family house, detached	1,047	57.19	58.23	490	72.93	73.27
	One-family house, attached	218	9.21	9.38	82	9.93	9.98
	Apartment (1–9 units)	302	10.40	10.59	60	6.98	7.01
	Apartment (≥ 10 units)	344	11.65	11.86	72	7.00	7.03
Year of construction	Missing	840	28.10	—	—	—	—
	1990–present	287	19.61	27.28	—	—	—
	1978–1989	265	14.84	20.64	—	—	—
	1960–1977	304	14.35	19.96	300	39.43	39.43
	1950–1959	168	7.43	10.34	158	19.38	19.38
	1940–1949	82	4.27	5.94	76	11.00	11.00
	Before 1940	209	11.39	15.84	197	30.19	30.19
Anyone smoke inside the home	Missing	23	1.50	—	1	0.46	—
	Yes	430	20.78	21.09	159	22.59	22.69
	No	1,702	77.73	78.91	571	76.95	77.31
Presence of deteriorated paint inside pre-1950 home[a]	Missing	239	7.87	—	0	—	—
	Yes	121	5.99	6.50	112	15.64	15.64
	No	1,795	86.14	93.50	619	84.36	84.36
Window, cabinet, or wall renovation in pre-1978 home[b]	Missing	176	6.02	—	9	0.64	—
	Yes	175	9.72	10.34	166	26.34	26.51

TABLE 3 *(Continued)*

			All homes			Pre-1978 homes	
			Weighted percent			Weighted percent	
	No	1,804	84.26	89.66	556	73.02	73.49
Window, cabinet, or wall renovation in pre-1950 home[c]	Missing	174	5.97	—	7	0.49	—
	Yes	70	3.98	4.23	65	10.69	10.74
	No	1,911	90.05	95.77	659	88.82	89.26
Race/ethnicity	Non-Hispanic white	618	57.09	57.09	252	64.14	64.14
	Non-Hispanic black	634	15.32	15.32	188	12.54	12.54
	Hispanicd	837	23.82	23.82	265	20.01	20.01
	Other	66	3.77	3.77	26	3.31	3.31
Country of birth	Missing	4	0.19	—	1	0.09	—
	United States	2,088	97.25	97.43	715	98.28	98.38
	Mexico	39	0.87	0.87	7	0.43	0.43
	Elsewhere	24	1.70	1.70	8	1.19	1.19

[a]Yes = presence of deteriorated paint inside AND pre-1950 home; no = no deteriorated paint inside OR post-1950 home.
[b]Yes = window, cabinet, or wall renovation AND pre-1978 home; no = no renovation OR post-1978.
[c]Yes = window, cabinet, or wall renovation AND pre-1950 home; no = no renovation OR post-1950.
[d]Sixty-six percent of Hispanics are Mexican Americans.

PbB Modeling Results

Although the models to predict log-PbB based on a cubic function of log floor PbD indicated that the cubic terms are not significant for two of the three data sets (the HUD Evaluation and Rochester), the quadratic terms were significant for all four data sets (Table 1). Figure 1 presents the predicted functions for the four data sets from the 5th to 95th floor PbD percentiles for each study except NHANES, which goes up to the 99.5th percentile (24.2 μg/ft^2). The figure shows that the slope and curvature of the relationship between log floor PbD and log PbB observed for the NHANES data are similar to other studies.

Children's PbB is strongly predicted by floor PbD and surface type and condition of floor (Table 4), with higher PbB associated with uncarpeted floors that were not smooth and not cleanable. Differences in the effect of PbD on PbB for uncarpeted smooth and cleanable, low-pile carpet and high-pile carpet were not significant, so these surfaces/conditions were combined. Natural log-transformed windowsill PbD, PIR, and age were also significant predictors of PbB.

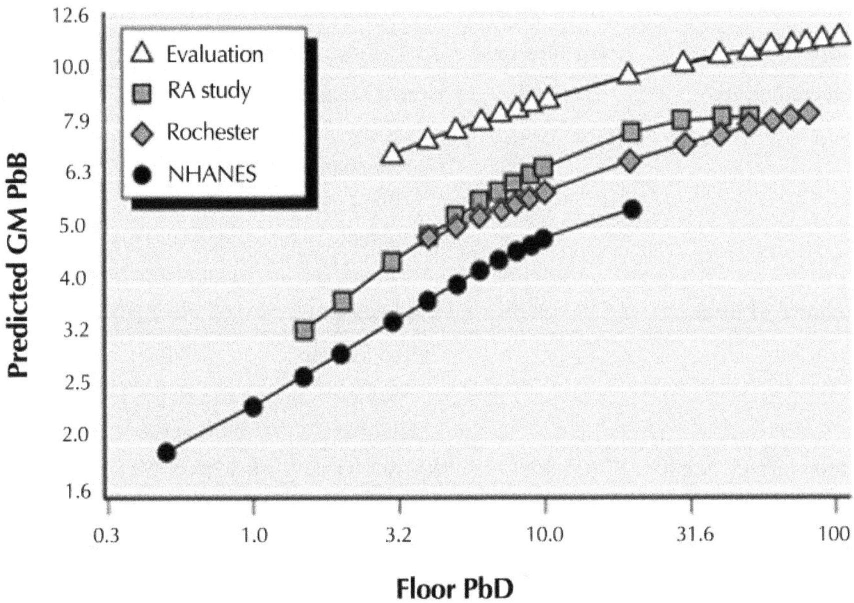

FIGURE 1 Predicted PbB (μg/dL) based on floor PbD (μg/ft2) by data set.

TABLE 4 Linear Model Results for Log Children's PbBa.

Variables	Overall p-value	Levels	Estimate (SE)	p-Value
Intercept	0.172		−0.517 (0.373)	0.172
Age (in years)	< 0.001	Age	2.620 (0.628)	< 0.001
		Age^2	−1.353 (0.354)	< 0.001
		Age^3	0.273 (0.083)	0.002
		Age^4	−0.019 (0.007)	0.008
Year of construction	0.014	Intercept for missing	−0.121 (0.052)	0.024
		1990 present	−0.198 (0.058)	0.001
		1978–1989	−0.196 (0.060)	0.002
		1960–1977	−0.174 (0.056)	0.003
		1950–1959	−0.207 (0.065)	0.003
		1940–1949	−0.012 (0.072)	0.870
		Before 1940	0.000	—
PIR	< 0.001	Intercept for missing	0.053 (0.065)	0.420
		Slope	−0.053 (0.012)	< 0.001

TABLE 4 *(Continued)*

Variables	Overall p-value	Levels	Estimate (SE)	p-Value
Race/ethnicity	< 0.001	Non-Hispanic white	0.000	—
		Non-Hispanic black	0.247 (0.035)	< 0.001
		Hispanic	−0.035 (0.030)	0.251
		Other	0.128 (0.070)	0.073
Country of birth	0.002	Missing	−0.077 (0.219)	0.728
		United Statesb	0.000	—
		Mexico	0.353 (0.097)	< 0.001
		Elsewhere	0.154 (0.121)	0.209
Floor surface/condition × log floor PbD	< 0.001	Intercept for missing	0.178 (0.094)	0.065
		Not smooth and cleanable	0.386 (0.089)	< 0.001
		Smooth and cleanable or carpeted	0.205 (0.032)	< 0.001
Floor surface/condition × $(\log \text{floor PbD})^2$		Not smooth and cleanable	0.023 (0.015)	0.124
		Smooth and cleanable or carpeted	0.027 (0.008)	0.001
Floor surface/condition × $(\log \text{floor PbD})^3$		Uncarpeted not smooth and cleanable	−0.020 (0.014)	0.159
		Smooth and cleanable or carpeted	−0.009 (0.004)	0.012
Log windowsill PbD	0.002	Intercept for missing	0.053 (0.040)	0.186
		Slope	0.041 (0.011)	< 0.001
Home-apartment type	< 0.001	Intercept for missing	−0.064 (0.097)	0.511
		Mobile home or trailer	0.127 (0.067)	0.066
		One family house, detached	−0.025 (0.046)	0.596
		One family house, attached	0.000	—
		Apartment (1–9 units)	0.069 (0.060)	0.256

TABLE 4 *(Continued)*

Variables	Overall p-value	Levels	Estimate (SE)	p-Value
		Apartment (≥ 10 units)	−0.133 (0.056)	0.022
Anyone smoke inside the home	0.015	Missing	0.138 (0.140)	0.331
		Yes	0.100 (0.040)	0.015
		No	0.000	—
Log cotinine concentration (ng/dL)	0.004	Intercept for missing	−0.150 (0.063)	0.023
		Slope	0.039 (0.012)	0.002
Window, cabinet, or wall renovation in a pre-1978 home	0.045	Missing	−0.008 (0.061)	0.896
		Yes	0.097 (0.047)	0.045
		No	0.000	—

[a]n = 2,155; R^2 = 40%.
[b]Includes the 50 states and the District of Columbia.

Non-Hispanic black children had significantly higher PbB than non-Hispanic whites ($p < 0.001$). Country of birth was also a significant predictor of PbB, with Mexican-born associated with higher PbB ($p = 0.003$). Children living in apartment buildings with ≥ 10 units were found to have lower PbB than children living in single-family detached or attached dwellings ($p = 0.005$ and $p = 0.022$, respectively). As expected, children living in newer housing have significantly lower PbB compared with children living in housing built before 1940 ($p < 0.001$). Children living in homes built before 1978 that had renovation (within the preceding 12 months), which often disturbs paint lead, had higher PbB ($p = 0.045$).

Children who resided in a home where smoking occurred inside had significantly higher PbB than children who lived in homes with no smoking ($p = 0.015$). Even after controlling for the presence of smoking in the linear model, increasing log cotinine concentrations were associated with increasing PbB ($p = 0.002$).

Table 5 presents the logistic regression results for predicting PbB ≥ 5 μg/dL and ≥ 10 μg/dL. If a variable was significant in one logistic regression model but not the other model, the cells for the variable contain a dash (—). Although most of the variables that were significant in the linear regression model were also significant in the 5 μg/ dL logistic regression model, the 10 μg/dL logistic regression model identified fewer significant predictors. The proportion of variability in the dependent variable accounted for by the model (R^2) for the 5 μg/dL and 10 μg/dL logistic models were much lower than for the linear model (16% and 5% vs. 40%, respectively). This result was attributable to the loss of information from using the dichotomous PbB outcomes in the logistic regression models and to the small number of children observed with PbB

≥ 10 μg/dL. The odds of having a PbB ≥ 5 μg/dL and ≥ 10 μg/ dL for non-Hispanic blacks were about twice those of non-Hispanic whites [odd ratio (OR) = 2.04 and 2.01, respectively]. The odds of a PbB ≥ 5 μg/dL for children born in Mexico were 11.69 times those of children born in the United States. However, country of birth was not a significant factor in predicting PbB ≥ 10 μg/dL. The odds of having a PbB ≥ 5 μg/dL were more than three times higher for children living in pre-1950 housing with renovation than for children living in other homes (OR = 3.33). The odds of having a PbB ≥ 10 μg/dL were more than three times higher for children living in pre-1950 housing with deteriorated paint inside than for children living in other homes (OR = 3.53).

TABLE 5 Model Results for Log Odds Children's PbB ≥ 5 μg/dL and ≥ 10 μg/dLa.

		PbB ≥ 5 μg/dL			PbB ≥ 10 μg/dL		
Term	**Levels**	**Overall *p*-value**	**Estimate (SE)**	***p*-Value**	**Overall p-value**	**Estimate (SE)**	***p*-Value**
Intercept		0.005	−13.004 (4.365)	0.005	0.048	−14.170 (6.976)	0.048
Age (in months)	Age	0.007	18.783 (7.069)	0.011	0.068	14.703 (11.140)	0.194
	Age^2		−10.455 (4.039)	0.013		−6.801 (6.673)	0.314
	Age^3		2.358 (0.959)	0.018		1.170 (1.687)	0.492
	Age^4		−0.189 (0.081)	0.024		−0.066 (0.149)	0.659
PIR	Intercept for missing	0.006	0.319 (0.444)	0.477	—	—	—
	Slope		−0.267 (0.099)	0.010		—	—
Race/ethnicity	Non-Hispanic white	0.003	0.000		0.038	0.000	
	Non-Hispanic black		0.712 (0.303)	0.023		0.696 (0.373)	0.068
	Hispanic		−0.468 (0.336)	0.171		−0.590 (0.513)	0.257
	Other		−0.048 (0.928)	0.959		−0.118 (1.002)	0.907
Country of birth	Intercept for missing	0.002	−0.518 (1.140)	0.652	—	—	—
	United Statesb		0.000	—	—	—	—

TABLE 5 *(Continued)*

		PbB ≥ 5 µg/dL			PbB ≥ 10 µg/dL		
Term	**Levels**	**Overall *p*-value**	**Estimate (SE)**	***p*-Value**	**Overall p-value**	**Estimate (SE)**	***p*-Value**
	Mexico		2.459 (0.641)	< 0.001	—	—	—
	Elsewhere		0.113 (1.145)	0.922	—	—	—
Log floor PbD	Intercept for missing	< 0.001	0.989 (0.410)	0.020	< 0.001	1.405 (0.630)	0.031
	Slope		0.807 (0.133)	< 0.001		0.710 (0.155)	< 0.001
Log window-sill PbD	Intercept for missing	0.056	0.466 (0.336)	0.172	0.071	1.234 (0.653)	0.066
	Slope		0.198 (0.080)	0.017		0.242 (0.102)	0.022
Home-apartment type	Intercept for missing	0.029	−0.434 (0.727)	0.553	0.048	1.638 (0.802)	0.047
	Mobile home or trailer		−0.078 (0.428)	0.857		0.480 (0.605)	0.432
	One-family house, detached		−0.373 (0.295)	0.214		0.212 (0.357)	0.556
	One-family house, attached		0.000	—		0.000	—
	Apartment (1–9 units)		−0.276 (0.361)	0.449		0.334 (0.508)	0.515
	Apartment (≥ 10 units)		−1.022 (0.326)	0.003		−1.173 (0.569)	0.045
Window, cabinet, or wall renovation in pre-1950 home	Missing	0.004	−0.052 (0.320)	0.872	—	—	—
	Yes		1.203 (0.399)	0.004	—	—	
	No		0.000	—		—	—
Presence of deteriorated paint inside a pre-1950 home	Intercept for missing	—	—	—	0.019	−0.012 (0.292)	0.968

TABLE 5 *(Continued)*

		PbB ≥ 5 μg/dL			PbB ≥ 10 μg/dL		
Term	**Levels**	**Overall *p*-value**	**Estimate (SE)**	***p*-Value**	**Overall p-value**	**Estimate (SE)**	***p*-Value**
	Yes		—	—	—	1.263 (0.520)	0.019
	No		—	—	—	0.000	—
Log cotinine concentration (ng/dL)	Intercept for missing	< 0.001	−0.299 (0.378)	0.434	0.006	−1.074 (0.931)	0.255
	Slope		0.483 (0.117)	< 0.001		0.455 (0.153)	0.005

[a]$n = 2{,}155$; $R^2 = 16\%$ and 5%. Approximate R^2 from Cox–Snell methodology.
[b]Includes the 50 states and the District of Columbia.

Floor PbD Thresholds

Table 6 presents the model predictions for average children living in a pre-1978 home for a range of floor PbD after controlling for the covariates described above. At a floor PbD of 6 μg/ft^2, the models predict that 2.7% of children have PbB ≥ 10 μg/dL and 16.5% have PbB ≥ 5 μg/ dL, and that the GM PbB is 3.4 μg/dL. When floor PbD is 12 μg/ft^2, the models predict that 4.6% of children have PbB ≥ 10 μg/dL and 26.8% have PbB ≥ 5 μg/dL, and that the GM PbB is 3.9 μg/dL. The upper bound of the 90% confidence interval (CI) for a prediction approximates the 95% upper bound for the prediction. For example, when floor PbD is 12 μg/ft^2, the 90% CI for the probability that PbB is ≥ 10 μg/dL is between 2.7 and 7.9%. This means that we are approximately 95% confident that the probability that PbB ≥ 10 μg/dL is < 7.9%. The information presented assumes that floor PbD is equal to the specified value. If floor PbD is less than the specified value, the predicted GM PbB and probabilities would be lower than those in Table 6.

TABLE 6 Estimated PbB for Children Living in pre-1978 Housing by Floor PbD, NHANES 1999–2004.

Floor PbD (μg/ft^2)	**Percent of homes ≥ floor PbD**	**GM PbB[a]**	**Probability (%) PbB ≥ 10 μg/dLb**	**Probability (%) PbB ≥ 5 μg/dLc**
0.25	79.1	1.7 (1.6–1.8)	0.2 (0.1–0.6)	1.1 (0.7–1.8)
0.50	55.4	1.9 (1.8–2.0)	0.4 (0.1–1.0)	2.1 (1.4–3.1)
1.00	30.5	2.2 (2.1–2.3)	0.6 (0.3–1.5)	3.8 (2.7–5.5)
1.50	21.8	2.4 (2.3–2.6)	0.9 (0.4–1.9)	5.4 (3.7–7.9)
2	16.7	2.6 (2.4–2.8)	1.1 (0.6–2.2)	6.9 (4.6–10.2)

TABLE 6 *(Continued)*

Floor PbD (µg/ft²)	Percent of homes ≥ floor PbD	GM PbB[a]	Probability (%) PbB ≥ 10 µg/dL[b]	Probability (%) PbB ≥ 5 µg/dL[c]
4	8.0	3.1 (2.8–3.4)	2.0 (1.1–3.5)	12.1 (7.7–18.5)
5	4.9	3.3 (2.9–3.6)	2.3 (1.3–4.1)	14.4 (9.0–22.2)
6	4.2	3.4 (3.0–3.8)	2.7 (1.5–4.7)	16.5 (10.2–25.6)
7	3.7	3.5 (3.1–4.0)	3.0 (1.7–5.3)	18.5 (11.3–28.7)
8	3.5	3.6 (3.2–4.1)	3.4 (2.0–5.8)	20.3 (12.3–31.7)
9	3.3	3.7 (3.3–4.2)	3.7 (2.1–6.4)	22.1 (13.3–34.4)
10	3.0	3.8 (3.3–4.3)	4.0 (2.3–6.9)	23.8 (14.2–36.9)
12	2.5	3.9 (3.4–4.5)	4.6 (2.7–7.9)	26.8 (16.0–41.5)
14	2.1	4.0 (3.5–4.7)	5.2 (3.0–9.0)	29.6 (17.5–45.5)
16	1.4	4.1 (3.6–4.8)	5.8 (3.3–10.0)	32.2 (19.0–49.0)
18	1.3	4.2 (3.6–4.9)	6.4 (3.6–11.0)	34.5 (20.3–52.1)
20	1.3	4.3 (3.6–5.0)	6.9 (3.9–11.9)	36.6 (21.6–54.9)
22	1.2	4.3 (3.7–5.1)	7.4 (4.1–12.9)	38.6 (22.8–57.3)
24	1.2	4.4 (3.7–5.2)	7.9 (4.4–13.8)	40.5 (23.9–59.6)
26	0.7	4.4 (3.7–5.2)	8.4 (4.6–14.8)	42.2 (25.0–61.6)
28	0.7	4.4 (3.7–5.3)	8.9 (4.8–15.7)	43.8 (26.0–63.5)
30	0.7	4.4 (3.7–5.4)	9.3 (5.1–16.6)	45.4 (26.9–65.2)
32	0.6	4.5 (3.7–5.4)	9.8 (5.3–17.4)	46.8 (27.9–66.7)
34	0.6	4.5 (3.7–5.5)	10.2 (5.5–18.3)	48.1 (28.7–68.1)
36	0.5	4.5 (3.6–5.6)	10.7 (5.7–19.1)	49.4 (29.6–69.4)
38	0.4	4.5 (3.6–5.6)	11.1 (5.9–20.0)	50.6 (30.4–70.6)
40	0.4	4.5 (3.6–5.7)	11.5 (6.1–20.8)	51.8 (31.2–71.8)

[a]90% CI based on the linear model for log PbB.
[b]90% CI based on the logistic model for PbB ≥ 10 µg/dL.
[c]90% CI based on the logistic model for PbB ≥ 5 µg/dL.

DISCUSSION

We found the GM PbB for children 12–60 months of age in the United States between 1999 and 2004 was 2 μg/dL and that 20 children per 1,000 had PbB ≥ 10 μg/dL. A prior study analyzing NHANES data collected 1994–1998 found that 63 children per 1,000 had PbB ≥ 10 μg/dL [33]. Our findings show that 81 children per thousand had PbB ≥ 5 μg/dL. Although there is a clear and significant decline over time in childhood lead exposure demonstrated by these prevalence estimates from NHANES, there is still an unacceptable number of children who are poisoned each year.

Age, race/ethnicity, PIR, and year of construction of housing all significantly predicted PbB of children, which is consistent with other studies [1, 34]. Prior studies also found that PbB is typically higher in African-American children than in white children [6, 29, 35] and is higher in children living in poverty and in older homes [1].

Previous studies using NHANES data have also documented the relationship between exposure to tobacco smoke and PbB [33, 36]. Similar to our finding that serum cotinine was associated with PbB ≥ 10 μg/dL, Mannino et al. [36] found that high levels of serum cotinine (a biomarker of exposure to environmental tobacco smoke) for older children 4–16 years of age was associated with PbB ≥ 10 μg/dL.

Prior studies have not demonstrated that children living in apartment buildings with ≥ 10 units are more likely to have lower PbB than children living in single-family detached houses. Although apartment buildings with ≥ 10 units tended to be of more recent construction than single-family detached homes and smaller apartment building (5%, 17%, and 78% constructed before 1940, respectively), not all the effect of home-apartment type is captured by the year of construction, because both variables are significant in the model. Although other studies suggest that lead hazards are more likely to be found in rental units than in owner-occupied properties [11], it is possible that owners of large apartment buildings may have more resources available for scheduled maintenance programs, which could help address lead hazards, compared with owners of smaller apartment buildings and single-family detached homes.

Despite having a relatively small number of children who were born outside the United States, our results indicate that Mexican-born was a strong predictor of PbB. A previous study examining the PbB of children living along the U.S.–Mexico border also found that children living in Mexico had higher PbB than children living in the United States. [37]. This finding may reflect continued use of lead-containing items imported from Mexico (e.g., pottery, foods, folk medicine) by families that recently resided there. Research has documented that use of these items can result in elevated PbB in children [38].

Additionally, our study supports the association between PbB and renovation and floor and sill PbD, as expected. Other studies have shown that renovation activities can influence floor PbD [13] and that floor PbD is a strong predictor of a child's PbB [8, 31, 39, 40]. The U.S. EPA [41] recently promulgated a regulation intended to control lead exposures from renovation.

The rate of change in PbB with respect to floor PbD levels observed in this most recent NHANES analysis is similar to that found in three other studies analyzed here: the Evaluation, the RA Study, and the Rochester Study (Table 1). These other data sets are from higher-risk populations and therefore have higher PbD and PbB levels. The similarities in the PbB/PbD slope in the different studies indicate that it is reasonable to use the NHANES data to make inferences at higher floor PbD and PbB.

The current federal floor PbD standard of 40 μg/ft^2 was established based on pre-1995 data from the Rochester Lead-in-Dust Study and a pooled analysis of 12 older epidemiologic studies using slightly different methods [8, 18, 26, 42]. The Rochester cohort and most of the studies comprising the pooled analysis were based on high-risk children and housing. The pooled analysis estimated that 95.3% of children 6–36 months of age would be protected from having a PbB ≥ 15 μg/ dL, using a floor PbD threshold of 40 μg/ ft^2 and holding other sources of lead to their respective national averages in the residential environment [8]. In the U.S. EPA analysis, the floor standard of 40 μg/ft^2 was established jointly with standards for lead in windowsill dust, soil, and interior paint to protect at least 95% of children 12–30 months of age from developing a PbB ≥ 10 μg/dL when the windowsill and soil lead standards were also met [18, 26]. Although the current 40 μg/ft^2 standard was based on protecting children from developing high PbB (i.e., PbB ≥ 10 μg/dL or ≥ 15 μg/dL), the importance of preventing lower childhood lead exposure is illustrated by research that has demonstrated significant lead-related IQ decrements in children with PbB < 10 μg/dL [3, 5]

A strength of our study is that we were able to show the relationship of a range of floor PbD levels on children's PbB while controlling for other significant predictors in a nationally representative sample of children. PbD and PbB from 1999 to 2004 were much lower than those observed in the earlier studies of higher-risk populations that were the foundation of the current floor PbD standard. In fact, these new data made the logistic model to predict PbB ≥ 10 μg/dL problematic, because only 2% of PbB (n = 51 of 2,155) were ≥ 10 μg/dL. Consequently, the percent of variation (R^2) explained by the predictors in the 10 μg/dL logistic model was much lower than that of the linear model (R^2 = 5% vs. R^2 = 40%). We present the logistic regression model for 5 μg/dL because no other PbB thresholds have regulatory significance, and 11% of children had PbB ≥ 5 μg/dL (237 of 2,155 children; R^2 = 16%). Iqbal et al. [43] suggest that the threshold for elevated PbB may be lowered from 10 to 5 μg/dL and examines the impact of this reduction.

NHANES collected both health and environmental data from a nationally representative sample of children between 12 and 60 months of age; however, the NHANES data are not necessarily representative of the U.S. housing stock. Iqbal and collaborators [43] found that for NHANES 1999–2002, a large number of children 1–5 years of age in NHANES (16.3%) had missing PbB values. Non-Hispanic white children, homeowners, and children from households with high-income levels and with health insurance had a higher percentage of missing PbB values. This may have inflated the estimates of GM PbB and overestimated the prevalence of PbB ≥ 5 μg/dL and PbB ≥ 10 μg/dL.

In addition, NHANES collected only a single floor PbD measurement in each house. Although the single measurement was from the room in which the children spent the most time, the average of several floor dust samples would likely provide a more precise estimate of a child's total exposure.

In this chapter we examined PbB across a range of floor PbD. An analysis of exposure pathways found that floor PbD has a direct effect on children's PbB, whereas sill PbD has an indirect effect on children's PbB as mediated by floor PbD [17]. In the NHANES data analyzed in this article, floor PbD is more predictive of PbB than sill PbD (R^2 = 19.4% for floors; R^2 = 11.9% for sills; R^2 = 23.0% for floors and sills combined). When floor PbD = 12 µg/ft², we show that 4.6% of children have PbB ≥ 10 µg/dL (Table 6). Based on the logistic model for 10 µg/dL, when floor PbD = 12 µg/ft², sill PbD = 90 µg/ft², and other covariates are at their national averages, the model predicts that 95% of children have PbB < 10 µg/dL. If homes have floor PbD below 12 µg/ft² and sill PbD below 90 µg/ft², less than 5% of children would have PbB ≥ 10 µg/dL.

The national estimate of the GM floor PbD in U.S. housing for 1998–2000 was 1.1 µg/ft² [11]. Furthermore, data from high-risk houses in the HUD evaluation study showed that PbD on floors continued to decline after the intervention, dropping from a GM of 14 µg/ft² immediately after intervention to a GM of only 4.8 µg/ft² 6 years after hazard control [31]. Together, these data demonstrate that floor PbD is well below the current federal standard of ≤ 40 µg/ft² for the vast majority of houses.

Historically, allowable PbD levels have declined as research has progressed. In the early 1990s, Maryland enacted a floor PbD standard of ≤ 200 µg/ft² [44]. U.S. EPA issued guidance in 1995 lowering the floor PbD level to ≤ 100 µg/ft², and in 1999–2001, HUD and U.S. EPA promulgated a floor PbD standard of ≤ 40 µg/ft², which has remained unchanged. Our findings suggest that floor and windowsill PbD should be kept as low as possible. Levels of PbD on floors between 6 µg/ft² and 12 µg/ft² can be expected to protect most children living in pre-1978 homes from having PbB ≥ 10 µg/dL. Protection at lower PbB would require lower PbD.

KEYWORDS

- **Centers for disease control and prevention**
- **Confidence interval**
- **Detection limit**
- **Lead exposure**
- **National Health and Nutrition Examination Survey**
- **Poverty-to-income ratio**

REFERENCES

1. CDC (Centers for Disease Control and Prevention). Blood lead levels—United States, 1999–2002. MMWR Morb Mortal Wkly Report 2005; 54:513–516.
2. National Research Council. Measuring lead exposure in infants, children and other sensitive populations. Washington, DC: National Academy Press; 1993.
3. Canfield RL, Henderson CR, Jr, Cory-Slechta DA, Cox C, Jusko TA, Lanphear BP. Intellectual impairment in children with blood lead concentrations below 10 microg per deci-liter. New England Journal of Medicine 2003; 348:1517–1526.
4. CDC. Preventing lead poisoning in young children: a statement by the centers for disease control. Report No. 99-2230. Atlanta, GA: Centers for Disease Control and Prevention; 1991.

5. Lanphear BP, Hornung R, Khoury J, Yolton K, Baghurst P, Bellinger DC, et al. Low-level environmental lead exposure and children's intellectual function: an international pooled analysis. Environmental Health Perspectives 2005; 113:894–899.
6. Lanphear BP, Hornung R, Ho M, Howard CR, Eberly S, Knauf K. Environmental lead exposure during early childhood. Journal of Pediatrics 2002; 140:40–47.
7. Levin R, Brown MJ, Kashtock ME, Jacobs DE, Whelan EA, Rodman J, et al. U.S. children's lead exposures, 2008: implications for prevention. Environmental Health Perspectives 2008; 116:1285–1293.
8. Lanphear BP, Matte TD, Rogers J, Clickner RP, Dietz B, Bornschein RL, et al. The contribution of lead-contaminated house dust and residential soil to children's blood lead levels. A pooled analysis of 12 epidemiologic studies. Environmental Resources 1998; 79:51–68.
9. Gibson L. A plea for painted railings and painted walls of rooms as the source of lead poisoning amongst Queensland children. Australia's Medicine Gazette 1904; 23:149–153.
10. Sayre JW, Charney E, Vostal J, Pless IB. House and hand dust as a potential source of childhood lead exposure. Archives of Pediatrics and Adolescent Medicine 1974; 127(2):167–170.
11. Jacobs DE, Clickner RP, Zhou JY, Viet SM, Marker DA, Rogers JW, et al. The prevalence of lead-based paint hazards in U.S. housing. Environmental Health Perspectives 2002; 110:A599–A606.
12. Dixon S, Wilson J, Galke W. Friction and impact surfaces: are they lead-based paint hazards? Journal of Occupational Environmental Hygiene 2007; 4:855–863.
13. Reissman DB, Matte TD, Gurnitz KL, Kaufmann RB, Leighton J. Is home renovation or repair a risk factor for exposure to lead among children residing in New York City? Journal of Urban Health 2002; 79:502–511.
14. Mielke HW. Lead in the inner cities. American Science 1999; 87:62–73.
15. Bornschein RL, Succop P, Kraft KM, Clark CS, Peace B, Hammond PB. Exterior surface lead, interior house dust lead and childhood exposure in an urban environment. In: Hemphill DD, editor. Trace substances in environmental health; Proceedings of University of Missouri's 20th Annual Conference. Columbia, MO: University of Missouri; 1987, 322–332.
16. Clark CS, Menrath W, Chen M, Succop P, Bornschein R, Galke W, et al. The influence of exterior dust and soil lead on interior dust lead levels that had undergone lead-based paint hazard control. Journal of Occupational Environmental Hygiene 2004; 1:273–282.
17. HUD. Evaluation of the HUD lead-based paint hazard control grant program: final report. Washington, DC: U.S. Department of Housing and Urban Development; 2004. [accessed 29 September 2008]. Available at: http://www.centerforhealthyhousing.org/HUD_National__Evaluation_Final_Report.pdf.
18. U.S EPA (U.S. Environmental Protection Agency). Identification of dangerous levels of lead. Final Rule. 40 CFR 745. Federal Register 2001; 66(4):1206.
19. NCHS (National Center for Health Statistics). National Health and Nutrition Examination Survey: NHANES 1999–2000; 2006a. Available at: http://www.cdc.gov/nchs/about/major/nhanes/nhanes99_00.htm [accessed 24 June 2008].
20. NCHS (National Center for Health Statistics). National Health and Nutrition Examination Survey: NHANES 2001–2002; 2006b. Available at: http://www.cdc.gov/nchs/about/major/nhanes/nhanes01-02.htm [accessed 24 June 2008].
21. NCHS (National Center for Health Statistics). National Health and Nutrition Examination Survey: NHANES 2003–2004; 2006c. Available at: http://www.cdc.gov/nchs/about/major/nhanes/nhanes2003-2004/nhanes03_04.htm [accessed 24 June 2008].
22. Office of Management and Budget. Statistical Policy Directive No. 14. Definition of poverty for statistical purposes; 1978. Available at : http://www.census.gov/hhes/www/povmeas/ombdir14.html [accessed 24 June 2008].
23. Dietrich JN, Ris MD, Succop PA, Berger OG, Bornschein RL. Early exposure to lead and juvenile delinquency. Neurotoxicol Teratology 2001; 23:511–518.

24. Tong SL, Baghurst PA, McMichael AJ, Sawyer MG, Mudge J. Lifetime exposure to environmental lead and children's intelligence at 11–13 years: the Port Pirie cohort study. BMJ 1996; 312:1569–1575.
25. Wasserman GA, Liu X, Lolacono NJ, Factor-Litvak P, Kline JK, Popovac D, et al. Lead exposure and intelligence in 7-year-old children: the Yugoslavia Prospective Study. Environmental Health Perspectives 1997; 105:956–962.
26. U.S. EPA, Office of Pollution Prevention and Toxics. Risk analysis to support standards for lead in paint, dust, and soil. EPA 747-R-97-006. Washington, DC: U.S. Environmental Protection Agency; 1998. Available at: http://www.epa.gov/lead/pubs/403risk.htm [accessed 24 June 2008].
27. U.S EPA. Lead; renovation, repair and painting. Final Rule. 40 CFR, Part 745. Federal Register 2008; 73(78):21692.
28. Galke W, Clark CS, Wilson J, Jacobs D, Succop P, Dixon S, et al. Evaluation of the HUD Lead Hazard Control Grantee Program: early overall findings. Environmental Resources 2001; 86(2):149–156.
29. Lanphear BP, Weitzman M, Eberly S. Racial differences in urban children's environmental exposures to lead. American Journal of Public Health 1996a; 86:1460–1463.
30. Lanphear BP, Weitzman M, Winter NL, Eberly S, Yakir B, Tanner M, et al. Lead-contaminated house dust and urban children's blood lead levels. American Journal of Public Health 1996b; 86(10):1416–1421.
31. Wilson J, Dixon S, Galke W, McLaine P. An investigation of dust lead sampling locations and children's blood lead levels. J Expo Sci Environ Epidemiol. 2007;17:2–12.
32. Wilson J, Pivetz T, Ashley P, Jacobs D, Strauss W, Menkedick J, et al. Evaluation of HUD-funded lead hazard control treatments at 6 years post-intervention. Environmental Resources 2006;102:237–248.
33. Teekens R, Koerts J. Some statistical implications of the log transformation of multiplicative models. Econometrica 1972; 40(5):793–819.
34. Bernard SM, McGeehin MA. Prevalence of blood lead levels ≥ 5 μg/dL among US children 1 to 5 years of age and socioeconomic and demographic factors associated with blood of lead levels 5 to 10 μg/dL, Third National Health and Nutrition Examination Survey, 1988–1994. Pediatrics 2003; 112:1308–1313.
35. Pirkle JL, Brody DJ, Gunter EW, Kramer RA, Paschal DC, Flegal KM, et al. The decline in blood lead levels in the United States. The National Health and Nutrition Examination Surveys (NHANES) JAMA 1994; 272:284–291.
36. Raymond JS, Anderson R, Feingold M, Homa D, Brown MJ. Risk for elevated blood lead levels in 3- and 4-year-old children. Maternal Child Health Journal; published online Oct 26, 2007.
37. Mannino DM, Albalak R, Grosse S, Repace J. Secondhand smoke exposure and blood lead levels in U.S. children. Epidemiology 2003; 14:719–727.
38. Cowan L, Esteban E, McElroy-Hart R, Kieszak S, Meyer PA, Rosales C, et al. Binational study of pediatric blood lead levels along the United States/Mexico border. International Journal of Hygiene and Environmental Health 2006; 209:235–240.
39. CDC. Preventing lead poisoning in young children. Atlanta, GA: National Center for Environmental Health and Injury Control; 1991.
40. Davies D, Thornton I, Watt J, Culbard E, Harvey P, Delves H, et al. Lead intake and blood lead in two-year-old UK urban children. Sci Total Environment 1990; 90:13–29.
41. Rabinowitz M, Leviton A, Bellinger D. Home refinishing, lead paint, and infant blood lead levels. American Journal of Public Health 1985; 75(4):403–404.
42. U.S EPA (U.S. Environmental Protection Agency). Lead; renovation, repair and painting. Final Rule. 40 CFR Part 745. Federal Register 2008; 73(78):21692.
43. HUD (U.S. Department of Housing and Urban Development). Requirements for notification, evaluation and reduction of lead-based paint hazards in federally owned residential property and housing receiving federal assistance. Final Rule. 24 CFR Part 35 Preamble. Federal Register 1999; p. 50181. Available at::/brgov.com/DEPT/ocd/pdf/1012_3final.pdf.

44. Iqbal S, Muntner P, Batuman V, Rabito FA. Estimated burden of blood lead levels ≥ 5 μg/dl in 1999–2002 and declines from 1988 to 1994. Environmental Resources 2008; 107:305–311.
45. Code of Maryland. Procedures for Abating Lead Containing Substances from Buildings. COMAR 26.02.07, Title 26, 8 August 1988. Baltimore, MD: Maryland Department of Environmental Regulations; 1988.

2 Teachers Working in PCB-Contaminated Schools

Robert F. Herrick, John D. Meeker, and Larisa Altshul

CONTENTS

INTRODUCTION

PCB contamination in the built environment may result from the release of PCBs from building materials. The significance of this contamination as a pathway of human exposure is not well characterized, however. This research compared the serum PCB concentrations, and congener profiles between 18 teachers in PCB–containing schools and referent populations.

Blood samples from 18 teachers in PCB–containing schools were analyzed for 57 PCB congeners. Serum PCB concentrations and congener patterns were compared between the teachers, to the 2003–2004 NHANES (National Health and Nutrition Examination Survey) data, and to data from 358 Greater Boston area men. Results: Teachers at one school had higher levels of lighter (PCB 6–74) congeners compared to teachers from other schools. PCB congener 47 contributed substantially to these elevated levels. Older teachers (ages 50–64) from all schools had higher total (sum of

33 congeners) serum PCB concentrations than age–comparable NHANES reference values. Comparing the teachers to the referent population of men from the Greater Boston area (all under age 51), no difference in total serum PCB levels was observed between the referents and teachers up to 50 years age. However, the teachers had significantly elevated serum concentrations of lighter congeners (PCB 6–74). This difference was confirmed by comparing the congener–specific ratios between groups, and principal component analysis showed that the relative contribution of lighter congeners differed between the teachers and the referents. Conclusions: These findings suggest that the teachers in the PCB–containing buildings had higher serum levels of lighter PCB congeners (PCB 6–74) than the referent populations. Examination of the patterns, as well as concentrations of individual PCB congeners in serum is essential to investigating the contributions from potential environmental sources of PCB exposure. Background Although PCBs have been banned from commerce since the late 1970s in most of the world, they persist in many environmental settings. Recent interest in building materials as possible sources of PCB contamination in the built environment has led to a series of investigations, and a set of EPA guidance documents on "PCBs in Caulk in Older Buildings" [1]. While evidence for contamination of indoor air from building materials has been growing recently, its significance as a possible source of PCB exposure to humans living or working in these buildings is less clear. Several investigations have provided evidence that occupancy of buildings containing PCB–rich building materials can result in elevated serum PCB levels. The investigation of indoor contamination resulting from PCB–containing caulking (sealant) material first reported by Benthe et al. [2] found elevated indoor air levels of the less–chlorinated PCB congeners 28, 52, and 101, however elevated serum levels were not observed among the building occupants. More detailed investigations confirmed the importance of building caulking and sealing materials as diffuse sources of PCB contamination [3, 4]. Gabio et al. [5], and Schwenk et al. [6] reported elevated PCB indoor air concentrations (congeners 28, 52 and 101) in schools containing PCB caulking materials. Teachers in these buildings had serum levels of PCB 28, 52, 101, 153, and 138 up to eight times greater than age–matched comparison subjects. Johansson et al. [7] reported a study of PCB–containing residential apartment buildings in which indoor air concentrations of PCB (sum of congeners 28, 52, 101, 138, 153, and 180) were found to be up to twice the levels measured in similar buildings without such sealants. Residents of the PCB–containing buildings showed significant elevations for serum levels of PCB 28, 74, 66, and 99, and also for total PCB based on the sum of 30 congeners. Liebel et al. [8], compared 377 children attending a PCB–containing school with 218 attending an uncontaminated school and reported that at least one of the lower chlorinated PCB congeners (of the six WHO–indicator congeners PCB 28, 52, 101, 138, 153, 180) could be detected in the blood of all the students at the PCB–containing school, compared to 27% of the students at the comparison school. Significantly higher median serum concentrations for PCBs 28, 52, and 101 were found in the students from the contaminated school, compared to the students from the control school. There was a significant positive association between years spent at the contaminated school and serum levels of the combined lower chlorinated congeners. The overall goal of the current study was to evaluate the possible relationship between serum PCB concentra-

tions, and employment in schools containing PCB in building materials. Specifically, our research aims included comparisons of total serum PCB concentrations between teachers who worked in these buildings and available reference groups who did not work in these buildings, as well as examination of the composition of the congener mix found in the serum of the teachers and the referents.

METHODS

A convenience sample of 18 teachers who worked in 3 schools containing PCB caulking material was recruited from current members of a local labor union. These three schools were constructed during the 1960s, corresponding to a period when PCB caulking was used in masonry buildings. The presence of PCB in the caulking of these schools was confirmed by testing conducted by the teachers' union. The subjects were volunteers recruited by the investigators from a population of approximately 3,000 members of the teachers' union. In addition to collecting a blood sample, the teachers were interviewed by the investigators (RH) about their work history, diet, smoking, and history of pregnancies and breastfeeding. A nonfasting venous blood sample was collected from the 18 volunteers on the last day of the school year in June 2009. Blood samples were centrifuged and serum stored in glass Wheaton vials at –20°C until analysis. Measurement of PCBs, p, p'–DDE and HCB in serum was conducted by the Organic Chemistry Analytical Laboratory, Harvard School of Public Health, Boston, MA. Target analytes included 57 individual PCB congeners, p, p'–DDE, and HCB. This set of target congeners included more low chlorinated (more volatile) congeners than are measured in the set of 35 NHANES congeners, because the methods used by our laboratory are applied to analyze air samples, as well as biological samples. Serum extracts were analyzed by dual capillary column gas chromatography with electron capture detection (GC/ micro ECD) and quantified based on the response factors of individual PCB congeners relative to an internal standard [9]. All final concentrations were reported after subtracting the amount in procedural blanks associated with the analytic batch. Results were not adjusted by surrogate recoveries. Wet weight serum PCB concentrations (e.g. ng/g serum), as well as lipid adjusted serum levels, were reported.

This study was approved by the Office of Human Research Administration at the Harvard School of Public Health.

Data Analysis

We used several approaches to examine the results of the serum analysis. First, we compared serum levels and congener patterns within the group of 18 teachers. We also compared the results from the teachers with the National Health and Nutrition Examination Survey (NHANES) participants. Our laboratory and the NHANES data report 33 of the PCB congeners in common. The primary difference between these laboratories is that of the 57 congeners our lab reports, 20 congeners are in the range of di–, tri– and tetrachloro congeners (IUPAC # 6–74), while the NHANES data reports only 6 in this range of lighter congeners.

The third comparison was between the teachers, and the results of serum PCB determinations from 358 men who were seeking infertility diagnosis from the Vincent

Burnham Andrology lab at Massachusetts General Hospital (MGH) in Boston (January 2000–May 2003). These are referred to as MGH referent samples [9]. These men were chosen as referents because their samples were run in the same lab as the teachers', using the same methods, their age ranges overlapped with the teachers, and they resided in the Greater Boston area (as did the teachers). The Harvard laboratory that analyzed the samples from the 18 teachers, and the MGH referents reported results for the same 57 PCB congeners.

We compared the serum levels of whole-weight and lipid adjusted congeners between the teachers and the referent groups, and examined the patterns of congeners between the groups. Finally, to further assess potential differences in congener patterns between groups, the composition of the congener mix in these 18 teachers was compared with the MGH referent subjects by principal components analysis (PCA). All measured congeners were normalized to PCB 153 and transformed by the natural logarithm prior to PCA analysis to obtain a unitless profile matrix as described previously [10].

RESULTS

Characteristics of the Teacher Study Subjects

The teachers ranged in age from 33 to 64 years. All had worked in their current school for at least 6 years, all were non-Hispanic whites, and the majority (13 of 18) were women.

Examining the trend in serum levels with age, we found that there was a strong correlation for the sum of wholeweight values of the 57 PCB congeners (Pearson correlations, ln–transformed PCB concentrations $R = 0.84$), and for the heavy PCBs (congeners 84–209, $R = 0.84$). The levels of lighter (congeners 6–74) PCBs were much less strongly correlated with age, $R = 0.49$. This age related trend was also observed for lipid–adjusted values (sum of 57 PCBs $R = 0.75$; heavy congeners $R = 0.77$, light congeners $R = 0.35$).

Within the group of 18 teachers, we observed distinct differences in the patterns of serum congener profiles. Seven of the 10 teachers from School A had serum levels of lighter (PCB 6–74) congeners that exceeded the median for the entire group of 18 (0.24 ng/g whole weight). These seven included several of the younger (< age 50) teachers who had some of the lowest total (sum of 57 congeners) serum levels. These lighter congeners, therefore, made a much greater contribution to the overall serum PCB levels for these subjects than for the other teachers (Table 1). For three of the younger teachers at School A, this elevated fraction of light congeners appeared to be driven by disproportionately high levels of PCB 47.

Comparison with 2003–2004 NHANES Data

The Harvard Organics Laboratory that analyzed the 18 teachers' samples, and the 2003–2004 NHANES reported serum levels for 33 of the same PCB congeners. The comparison of the 18 teachers with the age-stratified NHANES data only for the 33 congeners they report in common is presented in Table 2. Non-parametric statistical analysis comparing the sum of 33 congeners for the 18 teachers with the same age group (33–64 years) of NHANES subjects (Wilcoxon rank sum test) showed no difference between the groups overall (teacher median = 1.32 ng/g, NHANES median = 1.11 ng/g, p = 0.25), while the older teachers (age 50–64) had significantly higher levels than the age-comparable NHANES subjects (teacher median = 2.14 ng/g, NHANES median = 1.49 ng/g, p = 0.05). Teachers who reported more than one meal of dark fish or liver per week (subjects 5 and 18) had serum levels well above the NHANES GM for their age groups.

Examining the 6 congeners both laboratories reported in the range of lighter congeners (PCBs 28, 44, 49, 52, 66, 74), there was no significant difference (teacher median = 0.10 ng/g, NHANES median = 0.12 ng/g, p = 0.09) in serum concentrations. Comparing the percent contribution of the light congeners to the total (sum of 33 congeners) serum PCB level, the teachers' values were significantly lower than the NHANES values (teachers 8.2%, NHANES 11.3%, p = 0.005). These same patterns of differences were apparent when comparing lipid adjusted serum levels (Table 2).

Comparisons with MGH Referent Subjects

Comparing total serum PCB concentrations (sum of 57 congeners) between the 18 teachers with the 358 MGH referent subjects, 12 of the 18 exceeded the median level of total PCBs for the MGH referents, and 4 exceeded the upper 95% level for these referents. The oldest MGH referents were 51 years old, so we separately compared the 9 teachers younger than age 50 with the MGH referents. These younger teachers (under age 50) had total serum concentrations that were not different from the referents (teacher median = 1.01 ng/ g, MGH median = 1.09 ng/g, p = 0.61, Wilcoxon ranksum test). Comparing all 18 teachers with the MGH referents (Table 3), the teachers had somewhat higher total serum PCB levels overall (teacher median = 1.55 ng/g, MGH median = 1.09 ng/g, p = 0.02).

We also compared the concentrations of lighter congeners (di–, tri– and tetrachloro, PCB 6–74) in the 18 teachers with the 358 MGH referents. These include lesspersistent congeners that are more likely to volatilize from building materials than the heavier congeners (PCB 84–209). For the sum of congeners 6–74, all 18 teachers exceeded the MGH referent median, and 11 exceeded the upper 95% level for the MGH referents, including 5 of the 9 teachers aged 50 and younger. Statistical analysis by the Wilcoxon rank–sum test showed the teachers of comparable age to the MGH referents (50 and younger) to have significantly higher (teachers < 50 years median = 0.20 ng/g, MGH median = 0.08 ng/g, p < 0.0001) light PCB concentrations compared to the MGH referents. This significant difference was seen for the entire group of 18 as well (all teachers median = 0.23 ng/ g, MGH median = 0.08 ng/g, p < 0.0001). These same differences were found when comparing the lipidadjusted values as well.

The contribution of the lighter congeners (PCB 6–74) to the total serum level for each of the 18 teachers was compared to the MGH referent group. All 18 teachers exceeded the median % contribution of lighter congeners to the total for the MGH referents, by as much as a factor of approximately 5 (Table 1). The percent contribution of light congeners to the total serum PCB level was significantly higher for those aged 50 and younger (teachers < 50 years median = 18.7%, MGH median = 7.2%, p < 0.0001) as well as for the entire group of 18 teachers (all teachers median = 14.5%, MGH median = 7.2%, p < 0.0001).

TABLE 1 Teachers' Whole Weight Serum PCB Concentrations, % Light Congeners, and Congener Ratio.

Subject	Age	School	TotalPCB1	Light PCB2	Light PCB %total	PCB471	PCB47% total	Ratio PCB 47:153	
12	35	A	0.74	0.24	32.29		0.10	12.97	1.22
8	41	A	0.58	0.21	35.63		0.14	23.28	1.96
11	41	A	1.09	0.30	27.40		0.16	14.33	1.00
16	46	A	1.49	0.17	11.38		0.01	0.70	0.05
15	47	A	1.42	0.28	19.96		0.08	5.90	0.34
4	48	A	2.02	0.28	14.09		0.10	5.18	0.33
3	49	A	0.96	0.18	18.72		0.04	4.32	0.25
10	54	A	1.92	0.30	15.54		0.12	6.45	0.33
17	56	A	2.45	0.29	11.79		0.08	3.20	0.18
1	62	A	1.88	0.29	15.38		0.07	3.90	0.22
6	33	B	0.80	0.17	20.89		0.01	1.36	0.08
9	37	B	0.73	0.13	17.54		0.01	1.87	0.12
2	56	B	1.38	0.12	8.95		0.02	1.56	0.09
14	59	B	1.68	0.24	14.55		0.08	4.64	0.28

TABLE 1 *(Continued)*

Subject	Age	School	TotalPCB1	Light PCB2	Light PCB %to-tal	PCB471	PCB47% total	Ratio PCB 47:153	
5	52	C	2.46	0.20	8.11		0.03	1.02	0.06
7	60	C	2.17	0.20	9.17		0.02	1.04	0.05
13	62	C	4.37	0.40	9.20		0.02	0.46	0.02
18	64	C	5.32	0.71	13.29		0.03	0.05	0.03

Comparison of PCB Serum Congener Profiles

The composition of the mix of PCB congeners between the teachers and the referents was also examined. The first comparison of PCB congener profiles between teachers and MGH referents was to examine the ratios of the congener-specific group median concentrations. Because of the age differences between the 18 teachers and the MGH referents, we separated the 9 teachers who were under 50 from those over 50 and compared them with the MGH referents (Table 3). Median concentration ratios of 10 or greater between teachers less than age 50 and the MGH referents were found for two dichloro–PCBs (6 and 8), two trichloro–PCBs (33 and 37), one tetra–PCB (47), two penta–PCBs (87 and 97), two hexa–PCBs (136 and 149), and one hepta–PCB (174). The teacher to referent median ratios did not differ substantially between the group of 18 teachers, and the subgroup of 9 below the age of 50.

TABLE 2 Comparison of Teachers with Age–Stratified NHANES Levels for 33 PCB Congeners Common to Both Datasets, Non-Hispanic Whites.

Subject age	School	Σ33 NHANES congeners ng/g serum (pg/g lipid)	NHANES GM	NHANES 90%	NHANES95%
20–39	0.47(79.2)	1.03(157)	1.47(226)		
6	B	0.69(118)			
12	A	0.55(158)			
9	B	0.64(172)			
40–59	1.21(186)	2.48(375)	3.22(471)		
8	A	0.38(106)			
11	A	0.83(147)			
16	A	1.38(240)			
15	A	1.23(268)			
4	A	1.79(214)			
3	A	0.82(161)			
5	C	2.30(391)			
10	A	1.67(270)			

TABLE 2 *(Continued)*

Subject age	School	Σ33 NHANES congeners ng/g serum (pg/g lipid)	NHANES GM	NHANES 90%	NHANES95%
17	A	2.22(342)			
2	B	1.25(228)			
14	B	1.51(183)			
60+	2.27(347)	4.31(689)	5.91(929)		
7	C	2.03(364)			
1	A	1.66(268)			
13	C	4.16(694)			
18	C	4.98(1346)			

For 15 of the 18 teachers and all the MGH referents, the most prevalent congener found in serum was PCB 153. Calculating the ratios of the group median serum concentration for each congener to the group median concentration of congener 153 has the effect of normalizing the concentration of each PCB congener to the typically most prevalent congener [10]. For example, the ratio of the 18 teachers' median concentration of PCB congener 6 to the median concentration of congener 153 for the teachers can be compared to the same ratio in the MGH referents. A strength of this comparison is that it helps visualize the PCB composition for the subjects in each group, but it is insensitive to the absolute concentration of PCB in each sample. Comparing the congener-specific concentration ratios to PCB 153 between the 18 teachers and the MGH referents, substantial differences can be seen, a factor of 5 or more for PCBs 8, 33, 37, 41, 47, and 136. Another 13 congener ratios were at least twice as high in the 18 teachers compared to the referents. Most of these were in the range of lighter congeners (PCB 6–74, Figure 2). Three of the subjects had serum levels of congener 47 that equalled or exceeded their congener 153 levels (Table 1). These were among the younger teachers (age 41 and under), who all worked at School A. The specific congener to PCB 153 ratios for the 18 teachers and the MGH referents were very similar for the congeners above PCB 74 (penta–chlorinated and above, Figure 1).

TABLE 3 Teacher to NHANES and MGH Referent Median Ratios.

PCB	18 teacher	MGH referent	Ratio 18	Teachers	Ratio teachers	NHANES	RHAf
Congener	median	median	teachers	< 50 yrs median	<50 yrs:MGH	median	
	(ng/g serum)(ng/g serum)	MGH referent	(ng/g serum)	referents	subjects	<50 yrs/50 yrs (ng/g serum)	
6a	0.001	7 0.	0001	16.9931	0.0028	27.6006	0.001
8a, e	0.0071	0.0001	71.1125	0.0067	67.2269	0.001	
16 e	0.0038	0.00113	3.3367	0.0033	2.9499	0.005	
18 e	0.0033	0.00198	1.6906	0.0025	1.2732	0.003	
25a	0.0012	0.0001	12.1379	0.0005	5	0.024	
26a	0.0008	0.0001	7.7850	0.0008	8 0	.016	
28b, c, e	0.0222	0.00757	2.9357	0.0223	2.9417	0.031/0.030 0.12	
31 e	0.0023	0.00048	4.7290	0.0017	3.5014	0.019	
33a, b, e	0.0185	0.0001	185.483		0.01600	159.663	0.011
37a	0.0040	0.0001	40.4865	0.0025	25.2101	0.07	
41a	0.0016	0.0001	16.129	0.0009	8.7719	0.008	
44d, e	0.0034	0.00158	2.1274	0.0032	2.0416	0.013/0.012 0.009	
47a	0.0574	0.0068	8.4379	0.0836	12.2961	0.03	

TABLE 3 *(Continued)*

PCB	18 teacher	MGH referent	Ratio 18	Teachers	Ratio teachers	NHANES	RHAf
Congener	median	median	teachers	< 50 yrs median	<50 yrs:MGH	median	
	(ng/g serum)(ng/g serum)	MGH referent	(ng/g serum)	referents	subjects	<50 yrs/50 yrs (ng/g serum)	
49	0.0031	0.00057	5.5081	0.0023	4.0800	0.008/0.007 0.01	
52c, e	0.0054	0.00254	2.1201	0.0048	1.8748	0.002/0.017 0.009	
60b, d	0.0029	0.0024	1.1942	0.0026	1.0776	0.06	
66b, d, e	0.0159	0.0042	3.7751	0.0106	2.5284	0.008/0.009	0.07
70a, d, e	0.0070	0.00093	7.5341	0.0053	5.7094	0.05	
74b	0.0561	0.0331	1.6937	0.0364	1.1007	0.028/0.055	0.27
84	0.0009	0.00034	2.5577	0.0009	2.5355	0.04	
87a	0.0032	0.0001	31.7540	0.0031	30.9734	0.005/0.005	0.016
95e	0.0089	0.0042	2.1207	0.0103	2.4565	0.01	
97a	0.0052	0.0001	52.1779	0.0047	46.5116	0.06	
99	0.0444	0.0344	1.2923	0.0377	1.0959	0.024/0.036	0.28
101e	0.0173	0.0046	3.7594	0.0133	2.89271	0.010/0.009	0.03
105	0.0142	0.007	2.0222	0.0082	1.1699	0.006/0.010	0.32

TABLE 3 *(Continued)*

PCB	18 teacher	MGH referent	Ratio 18	Teachers	Ratio teachers	NHANES	RHAf
Congener	median	median	teachers	< 50 yrs median	<50 yrs:MGH	median	
	(ng/g serum)(ng/g serum)	MGH referent	(ng/g serum)	referents	subjects	<50 yrs/50 yrs (ng/g serum)	
110d	0.0081	0.0035	2.3041	0.0078	2.2167	0.007/0.007	0.014
118	0.0795	0.0595	1.3369	0.0613	1.0301	0.031/0.059	0.5
128	0.0029	0.0008	3.6754	0.0025	3.1512	0.001/0.001	0.3
135	0.0042	0.0015	2.7809	0.0043	2.8736	0.02	
136a	0.0043	0.0001	43.1323	0.00420	42.0168	0.008	
138	0.1527	0.158	0.9666	0.0798	0.5053	0.097/0.178	0.6
141	0.0024	0.007	0.3456	0.0018	0.2506	0.06	
146	0.0353	0.0215	1.6404	0.0189	0.8794	0.013/0.027	1.6
149a	0.0133	0.00016	83.329	0.0142	88.4956	0.004/0.004	0.04
151	0.0073	0.0015	4.8602	0.0070	4.6784	0.002/0.002	0.03
153	0.2611	0.2016	1.2953	0.1561	0.7745	0.132/0.251	1.1
156	0.0407	0.0272	1.4973	0.0140	0.5156	0.022/0.044	2.3
157	0.0109	0.0084	1.2961	0.0061	0.7310	0.005/0.010	1.1

TABLE 3 *(Continued)*

PCB	18 teacher	MGH referent	Ratio 18	Teachers	Ratio teachers	NHANES	RHAf
Congener	median	median	teachers	< 50 yrs median	<50 yrs:MGH	median	
	(ng/g serum)(ng/g serum)	MGH referent	(ng/g serum)	referents	subjects	<50 yrs/50 yrs (ng/g serum)	
167	0.0121	0.0074	1.6395	0.0079	1.0669	0.004/0.009	1.2
170	0.0806	0.06	1.3430	0.0368	0.6140	0.042/0/081	2
171	0.0106	0.0061	1.7353	0.0059	0.9643	0.7	
174a	0.0078	0.0001	77.771	0.0071	70.7965	0.024	
180	0.2278	0.142	1.6045	0.1067	0.7516	0.117/0.229	2.9
183	0.0280	0.0172	1.6255	0.0151	0.8788	0.010/0.019	0.5
187	0.0595	0.0445	1.3362	0.0293	0.6587	0.028/0.057	1.4
189	0.0036	0.0029	1.2403	0.0009	0.3025	0.001/0.001	2.6
194	0.0633	0.03	2.1090	0.0198	0.6583	0.026/0.056	3.1
195	0.0104	0.007	1.4849	0.0042	0.6002	0.006/0.012	1.1
196	0.0329	0.016	2.0590	0.0176	1.1029	0.022/0.042	0.8
199	0.0442	0.031	1.4271	0.0167	0.5376	0.023/0.053	0.04
203	0.0399	0.016	2.4930	0.0146	0.9159	1.8	

TABLE 3 *(Continued)*

PCB	18 teacher	MGH referent	Ratio 18	Teachers	Ratio teachers	NHANES	RHAf
Congener	median	median	teachers	< 50 yrs median	<50 yrs:MGH	median	
	(ng/g serum)(ng/g serum)	MGH referent	(ng/g serum)	referents	subjects	<50 yrs/50 yrs (ng/g serum)	
206	0.0257	0.016	1.6070	0.0113	0.7090	0.014/0.032	2.1
209	0.0086	0.0067	1.2903	0.0043	0.6433	0.007/0.020	1.8
201/177	0.01210	0.0093	1.3007	0.0101	1.0843	0.008/0.016	2.9/1

Legend: Superscripts refer to congeners that have been identified as markers of occupational or environmental, non–dietary PCB exposures. Congeners included in the NHANES data are in bold font.
Congeners identified as markers of non–dietary exposure.
aHerrick, et al., 2007.
bLoutano, et al., 1991.
cKontas, et al., 2004.
dWingfors, et al., 2008.
eDeCaprio, et al., 2005.
fRelative Human Accumulation factors, Brown, 1992 (Ref 13).

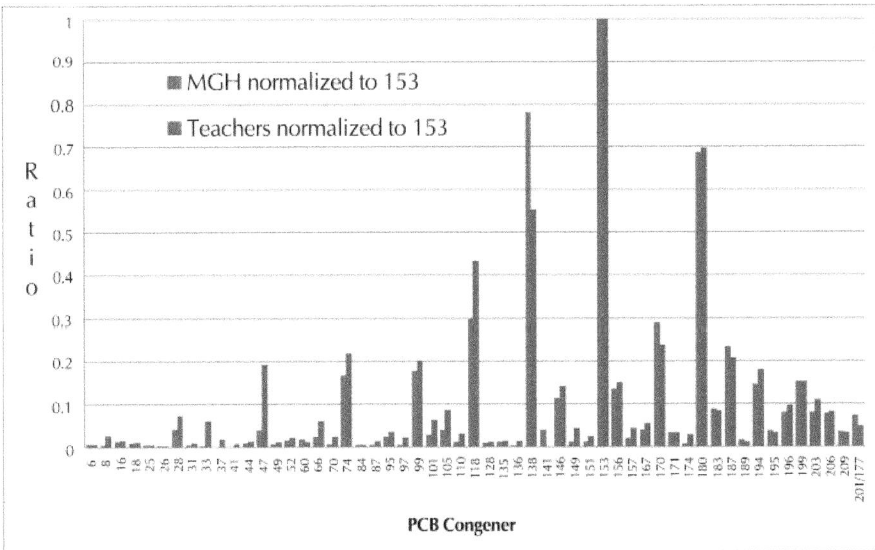

FIGURE 1 Comparison of median serum congener concentrations normalized to congener 153 concentration between teachers and MGH referents. Comparison of the height of the two bars (teachers and referents) for each congener illustrates the differences in the relative abundance of each congener in the serum of the teachers and the referents. The greatest differences are apparent in the lighter congeners, PCB 6–74, where the teachers' values consistently exceed the referents.

Principal Components Analysis

Principal components analysis (PCA) was used to further assess the congener composition patterns among the 18 teachers and the MGH referents. In examining the summary table, 33 of the 57 congeners loaded positively on principal components (PC) 1 and 2, including all the congeners lower than PCB 74. The differences between the component scores for the teachers and the MGH referents for PC1 and 2 were highly significant ($p < 0.0001$) by Wilcoxon rank-sum test. The score plot of the first two components shows a distinct difference between the populations, where the 18 teachers tend to have higher positive scores for both PC1 and PC2 (Figure 2).

DISCUSSION

Examining the pattern of congeners within the group of 18 teachers, there appears to be a relatively greater abundance of the lighter (PCB 6–74) congeners among the teachers from School A compared to the other two schools. The highest level of PCB 47 among teachers at schools B and C was 0.0800 ng/g, while five of the 10 teachers at School A exceeded this level. Details on the composition of each subjects' congener profile are presented in Additional File 4: Summary table of serum PCB whole weight (ng/g) by congener homolog group. All 18 teachers exceeded the median value of 0.0068 ng/ g for the MGH referents, however. PCB 47 (2, 2', 4, 4'–tetrachlorobiphenyl, CAS 2437–79–8) was reportedly present (0.07–0.14%) in the commercial formulation Aroclor 1254, which was commonly used as a plasticizer in caulking materials

[11]. PCB 47 is among the more volatile congeners, vapor pressure 8.63° 10–5 mm Hg at 25 deg C [12], and it has a relative human accumulation factor of 0.03 [13] indicating that is among the less persistent PCB congeners. Its presence among the teachers may reflect their recent exposure, as estimated human half-life values for 0.2 to 5.5 years have been reported [11] for PCB 47. By comparison, the estimated half-life for PCB 153 ranges from 0.9 to 47 years.

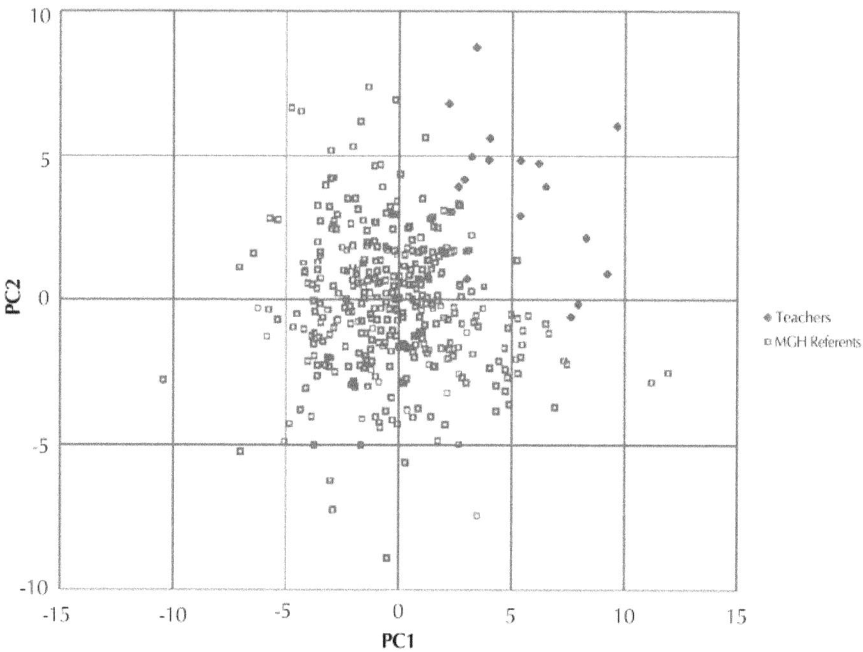

FIGURE 2 PCA Score plot components 1 and 2 for teachers and MGH referents.

In the case of the 18 teachers in the current study, there appeared to be a consistent difference in serum levels between the teachers and the MGH referents over the range of comparisons (Table 3). The difference is most apparent among congeners 6–74 (below the penta-chlorinated congeners). The relative human accumulation (RHA) [13] factors for these congeners are typically below 0.05, suggesting that they are more readily metabolized than the more persistent congeners (PCB 153 has a RHA of 1.1, for example). These are also the congeners with vapor pressures in the range of 10–4 to 10–8 mm Hg [tri- to hepta-chlorinated PCBs] that are sufficiently volatile to exist in the aerosol particle adsorbed and vapor phase [12]. This pattern of elevated serum congener levels may be attributed to the contribution of environmental sources, other than diet, to the overall PCB body burden in these teachers. The serum levels among the 18 teachers were elevated for all the congeners that have been identified as markers of non-dietary exposure, but only for one of the congeners higher than PCB 153 (Table 3). The ratios between the median congener levels among the teachers and the MGH referents were most prominently elevated among the lighter (PCB 6–74) congeners, but ratios of the measured dioxin-like PCB congeners (105, 118, 156, 157, 167, and 189) were all below 2, suggesting that occupancy of PCB-containing buildings did not substantially increase serum concentrations of these congeners.

Three of the men among the MGH referents reported that their occupation was "teacher," they were all aged 36 to 37 years. Their median total serum PCB was 2.69 ng/g (median MGH referents aged 30–39, 0.99 ng/g), and their median light congener level was 0.23 ng/g (MGH referents median 0.08 ng/g).

The PCA model further demonstrated significant differences between teachers and referents with regard to congener composition. The congeners that loaded most positively on PC1 and PC2 were PCB 6, 8, 16, 18, 25, 26, 31, 33, 37, 41, 44, 47, 49, 52, 66, and 95, confirming the differences observed in the other comparisons of the teachers and the MGH referents. Examining the score plot for the first two components, teachers had uniformly higher positive scores than the MGH referents. A challenge in comparing serum PCB levels between groups is the selection of PCB congeners to use as exposure metrics [10, 14, 15]. The choice of congeners for comparison is limited by the fact that reference data frequently reports levels only for the most abundant congeners found in serum. For example, PCBs 28, 52, 101, 118, 138, 153 and 180 are frequently measured in serum to assess exposure. Grandjean et al. [16] chose congeners based upon their detectability in chemical analyses (PCBs 118, 138, 153, 170, 180, and 187), and PCBs 105 and 156 as they are mono–ortho congeners of toxicological interest. These sets of PCBs include several highly chlorinated congeners without vicinal hydrogens that tend to be highly persistent, and accumulate in the environment [17–19]. The major source of exposure to these congeners appears to be diet, particularly the consumption of fatty fish [20, 21].

The World Health Organization (WHO) uses a set of 6 congeners as PCB indicators (28, 52, 101, 138, 153, and 180). Of these, 28 and 52 can be markers of indoor air contamination, while the others primarily have dietary sources [22]. For studies in the United States, the NHANES set of 35 PCB congeners is frequently used as a source of reference values. These congeners include those most frequently detected in the sampled population (138, 153, 180), and the set of 35 is weighted toward the hexa–chlorinated congeners and above (23 of the 35 are PCB 128 and higher).

Investigators [8, 23–26] have proposed sets of specific PCB congeners that may serve as markers to distinguish between PCB exposures primarily from dietary sources, and the contribution of other exposure sources, such as employment in capacitor plants, where exposure occurred both by inhalation and dermal contact. In cases where inhalation of PCBs volatilized from contaminated outdoor sites or occupancy of contaminated buildings is likely to be a significant exposure pathway, lighter, lower chlorinated congeners (< PCB 74, tetra-chloro and below) would be expected to be more characteristic markers of exposure. Comparison of these sets of markers may also provide information about the temporality of exposure by separating more recent exposures from earlier exposures, as the presence of less-persistent, more readily metabolizable congeners suggests recent contact with sources of exposure. Wolff et al.

[27] identified PCB congeners 28, 74, 118, 105, and 156 as unique markers of occupational exposure in a study of former capacitor plant workers. This was based upon the observation that levels of these congeners are low in the general population, and they are derived almost exclusively from the Aroclor mixtures used in the capacitor plants. De Caprio et al. [14] used polytopic vector analysis (PVA) to identify congener patterns in serum in a study of Akwesasne Mohawks with historical PCB exposure. Congeners 8, 18, 32/16, 31, 28, 33, 52, 44, 70, 66, 95, and 90/101 (primarily tri- and tetrachloro PCBs) were hypothesized to reflect recent inhalation exposure. Freels et al. [15] suggested that combinations of congener levels and their relative proportions should be considered relevant in tracking the source and pathway of PCB exposures. A recent animal study demonstrated that inhalation is a very efficient route of exposure and uptake, particularly for lower–chlorinated congeners [28].

We previously conducted a study comparing the congener-specific serum PCB concentrations between workers who removed old PCB caulk and the MGH referents. There were substantial differences between subject and referent mean and median serum levels for PCB 6, 16, 26, 33, 37, 41, 70, 97, and 136 [29]. For these congeners, the subject mean (and median) exceeded the reference mean (and median) by a factor of 5 or more. The congeners found to be most substantially elevated among the construction workers were those rapidly eliminated from the body, typically with relative human accumulation (RHA) factors from 0.001 to 0.07 [13] and short apparent human half-lives, e.g. 0.02 years, approximately 1 week for PCB 33 [11]. Notably, the NHANES data does not include any of these congeners (6, 16, 26, 33, 37, 41, 70, 97, and 136).

While our study provides evidence that the 18 teachers have elevated serum levels of certain light PCB congeners, suggesting an environmental source, several limitations must be considered in interpreting the data. First, the comparisons with the NHANES data are restricted by the dissimilarity in the sets of congeners reported by the two laboratories and differences in the laboratory methods used. The main reason for this dissimilarity apparently is that the analytical strategy of the CDC laboratory is designed for serum analysis, where a high number of light, lower chlorinated congeners would not be expected. In contrast, our laboratory chose the list of target congeners considering that it would be conducting analysis for a number of studies, some of which would include air samples, as well as biological samples. Therefore more low-chlorinated (more volatile) congeners were included in our analytical strategy. This mixture of congeners is examined in all matrices, including air samples and blood serum. Another limitation was that the MGH study included only men, while the majority (13 of 18) of our teachers were women. The 2003–2004 NHANES data indicated that total (sum of 35 congeners) levels were slightly higher in men than in women [30], so the difference we saw between the teachers and the MGH referents may underestimate the true difference. In addition, we cannot rule out the possibility that there are differences between men and women in elimination (biotransformation) of lower molecular weight PCBs. We were unable to recruit a referent group of teachers who worked only in schools built after 1980, which would be expected to be free of PCB-containing building materials. Comparing the MGH referents with NHANES data for the same age strata (20 to 51) for the 33 congeners reported in common, the MGH subjects have significantly elevated levels for the sum of the 33 congeners compared to the NHANES, however the levels of 5 of the 6 congeners in the PCB 6–74 range are significantly lower for the MGH subjects than the age–comparable NHANES values. Both the 18 teachers and the MGH referents have lower levels of the 6 light congeners than age-comparable NHANES results. While these levels may actually be lower, it is also possible that inter-laboratory differences may contribute to these findings.

Another factor we considered in the comparison between the teachers and the MGH referents is the lack of information we have about the contact the referents may have had with non-dietary sources of PCBs. The 18 teachers were all sampled at the end of the last day of school in 2009, but the MGH referents were sampled during weekday clinic visits, and we have no information, for example, on how recently or how much time they spent in a PCB-containing building. Finally, ours was a pilot study, limited by the small sample size (18 teachers) and the fact that we did not have access to characterize the indoor environments of the schools by air sampling. We cannot explain, for example, the differences in serum levels of light congeners (e.g., PCB 47) seen between the teachers at school A and the teachers from other schools. A larger study that includes a comprehensive evaluation of potential PCB sources associated with the buildings (caulk, fluorescent light ballasts, ceiling tiles, carpet adhesives, and paints) is clearly needed.

CONCLUSIONS

Despite the limitations, this study provides evidence that occupants of PCB-containing buildings have elevated serum levels of several PCB congeners. The nature of the difference between the teachers and the referents is not apparent by comparison with the NHANES reference values, which are weighted toward measuring the more persistent, highly chlorinated congeners. The congener profile for the 18 teachers, however, is significantly enriched with lighter congeners, including those that have been previously identified as coming from non-dietary sources. These findings show the importance of analyzing a wide range of PCB congeners when seeking to identify the possible sources of human exposures. In view of the growing concern about occupancy of PCB-containing buildings, investigators should assess as full a profile of congeners as feasible in environmental and biological samples to fully characterize the nature and sources of PCB body burdens.

KEYWORDS

- **Lighter congeners**
- **Principal components analysis**
- **Relative human accumulation**
- **World Health Organization**
- **Polytopic vector analysis**

REFERENCES

1. USEPA [United States Environmental Protection Agency]. PCBs in Caulk in Older Buildings. Washington, DC; 2009. Available at: http://www.epa.gov/pcbsincaulk/ [accessed 6 November 2010].
2. Benthe C, Heinzow B, Jessen H, Mohr S, Rotard W. Polychlorinated Biphenyls. Indoor air contamination due to Thiokol-rubber sealants in an office building. Chemosphere 1992; 25:1481–1486.
3. Kohler M, Tremp J, Zennegg M, Seiler C, Minder-Kohler S, Beck M, Lienemann P, Wegmann L, Schmi P. Joint sealants: an overlooked diffuse source of polychlorinated biphenyls in buildings. Environ Sci Technol 2005; 39:1967–1973.
4. Harrad S, Hazrati S, Ibarra C. Concentrations of polychlorinated biphenyls in indoor air and polybrominated diphenyl ethers in indoor air and dust in Birmingham, United Kingdom: implications for human exposure. Environ Sci Technol 2006; 40:4633–4638.
5. Gabrio T, Piechotowski I, Wallenhorst T, Klett M, Cott L, Friebel P, Link B, Schwenk M. PCB–blood levels in teachers, working in PCB–contaminated schools. Chemosphere 2000; 40:1055–1062.
6. Schwenk M, Gabrio T, Papke O, Wallenhorst T. Human biomonitoring of polychlorinated biphenyls and polychlorinated dibenzodioxins and dibenzofuranes in teachers working in a PCB-contaminated school. Chemosphere 2002; 47(2):229–233.
7. Johansson N, Hanberg A, Bergek S, Tysklind M. PCB in sealants are influencing the levels in indoor air. Organohalogen Compounds 2001; 52:436–439.
8. Liebel B, Schettgen T, Herscher G, Broding H-C, Otto A, Angerer J, Drexler H. Evidence for increased internal exposure to lower chlorinated polychlorinated biphenyls (PCB) in pupils attending a contaminated school. Int J Hyg Environ Health 2004; 23:315–324.
9. Hauser R, Chen Z, Pothier L, Ryan L, Altshul L. The relationship between human semen parameters and environmental exposure to polychlorinated biphenyls and p, p'–DDE. Environmental Health Perspectives 2003; 111:1505–1511.
10. Wingfors H, Selde AI, Nilsson C, Haglund P. Identification of markers for PCB exposure in plasma from Swedish construction workers removing old elastic sealants. Ann Occup Hyg 2006; 50:65–73.
11. ATSDR (Agency for Toxic Substances and Disease Registry). Toxicological profile for polychlorinated biphenyls. Atlanta; GA: U.S. Department of Health and Human Services; 2000.
12. Eisenreich SJ, Baker JE, Franz T, Swanson M, Rapaport RA, Strachan WMJ, Hites RA. Atmospheric deposition of hydrophobic organic contaminants to the Laurentian Great Lakes. In Fate of pesticides and chemicals in the environment. Edited by: Schnoor JL. New York, NY: John Wiley 1992; 51–78.
13. Brown JF. Determination of PCB metabolic, excretion, and accumulation rates for use as indicators of biological response and relative risk. Environ Sci Technol 1994; 28:2295–2305.
14. De Caprio AP, Johnson GW, Tarbell AM, Carpenter DL, Chiarenzelli JR, Morse GS, Santiago–Rivera AL, Schymura MJ. Akwesasne Task Force on the Environment: Polychlorinated biphenyl (PB) exposure assessment by multivariate statistical analysis of serum congener profiles in an adult Native American population. Environ Res 2005; 98:284–302.
15. Freeles S, Kaatz Chary L, Turyk M, Piorkowski J, Mallin K, Dimos J, Anderson H, McCann K, Burse V, Persky V. Congener profiles of occupational PCB exposure versus PCB exposure from fish consumption. Chemosphere 2007; 69:435–443.
16. Grandjean P, Budtz–Jørgensen E, Barr DB, Needham LL, Weihe P, Heinzow B. Elimination Half–lives of Polychlorinated Biphenyl Congeners in Children. Environmental Science Technology 2008; 42:6991–6996.
17. Borlakoglu JT, Wilkins JPG. Correlations between the molecular structures of polyhalogenated biphenyls and their metabolism by hepatic microsomal monooxygenases. Comp Biochem Physiol 1993; 105C:113–117.

18. Niimi AJ, Oliver BG. Biological half-lives of polychlorinated biphenyl [PCB] congeners in whole fish and muscle of Rainbow Trout [Salmo gairdneri]. Can J Fish Aquat Sci 1983; 40:1388–1394.
19. Andersson PL, Berg AH, Bjerselius R, Norrgren L, Olsén H, Olsson PE, Orn S, Tysklind M. Bioaccumulation of selected PCBs in Zebrafish; three-spined Stickleback; and arctic Char after three different routes of exposure. Arch Environ Contam Toxicol 2001; 40:519–530.
20. Alcock RE, Behnisch PA, Jones KC, Hagenmaier H. Dioxin-like PCBs in the environment–human exposure and the significance of sources. Chemosphere 1998; 37:1457–1472.
21. Schlummer M, Moser GA, Mclachlan MS. Digestive tract absorption of PCDD/Fs; PCBs; and HCB in humans: mass balances and mechanistic considerations. Toxicol Appl Pharmacol 1998; 152:128–137.
22. Broding HC, Schettgen T, Goen T, Angerer J, Drexler H. Development and verification of a toxicokinetic model of polychlorinated biphenyl elimination in persons working in a contaminated building. Chemosphere 2007; 68:1427–1434.
23. Luotamo M. Congener specific assessment of human exposure to polychlorinated biphenyls. Chemosphere 1991; 23:1685–1698.
24. Korrick SA, Altshul LM, Tolbert PE, Burse VW, Needham LL, Monson RR. Measurement of PCBs; DDE; and Hexachlorobenzene in Cord Blood from Infants Born in Towns; Adjacent to a PCB–Contaminated Waste Site. Journal of Exposure Analysis and Environmental Epidemiolog 2000; 10:743–754.
25. Kontsas H, Pekari K, Riala R, Back B, Rantio T, Priha E. Worker exposure to polychlorinated biphenyls in elastic polymer sealant renovation. Ann Occup Hyg 2004; 48:51–55.
26. Wolff MS, Fischbein A, Selikoff IJ. Changes in PCB serum concentrations among capacitor manufacturing workers. Environmental Research 1992; 59:202–216.
27. Wolff MS, Camann D, Gammon M, Stellman SD. Proposed PCB congener groupings for epidemiological studies. Environmental Health Perspectives 1997; 105:13–14.
28. Hu X, Adamcakova-Dodda, Lehmler H-J, Hu D, Kania-Korwel I, Hornbuckle KC, Thorne PS. Time course of congener uptake and elimination in rats after short-term inhalation exposure to an airborne polychlorinated biphenyl(PCB) mixture. Environmental Science Technology 2010; 44:6893–6900.
29. Herrick RF, Meeker JD, Hauser R, Altshul L, Weymouth GA. Serum PCB levels and congener profiles among US construction workers. Environmental Health 2007; 31:25.
30. Patterson DG, Wong L–Y, Wayman E, Turner SP, Caudill ES, DiPietro ES, McClure PC, Cash TP, Osterloh JD, Pirkle JL, Sampson EJ, Needham LL. Levels in the U.S. Population of those Persistent Organic Pollutants [2003–2004]. Included in the Stockholm Convention or in other Long-Range Transboundary Air Pollution Agreements. Environmental Science Technology 2009; 43:1211–1218.

3 Flame-Retardants' Effect on Hormone Levels and Semen Quality

John D. Meeker and Heather M. Stapleton

CONTENTS

INTRODUCTION

Organophosphate (OP) compounds, such as tris(1, 3-dichloro-2-propyl) phosphate (TDCPP) and triphenyl phosphate (TPP), are commonly used as additive flame-retardants and plasticizers in a wide range of materials. Although widespread human exposure to OP flame-retardants is likely, there is a lack of human and animal data on potential health effects.

Several chemicals commonly encountered in the environment have been associated with altered endocrine function in animals and humans, and exposure to some endocrine-disrupting chemicals may result in adverse effects on reproduction, fetal/child development, metabolism, neurologic function, and other vital processes [1]. Recent attention to the potential risks that environmental chemicals may pose to reproductive and developmental health has also been driven by reports of temporal downward trends in semen quality [2, 3] and male testosterone levels [4, 5]; increased rates of development anomalies of the reproductive tract, specifically hypospadias and cryptorchidism [6]; and increased rates of testicular cancer [7–9]. Public and scientific concern also stems from recent reports of inexplicable increases in the rates of thyroid cancer [10, 11], congenital hypothyroidism [12], and neurologic development disorders such as autism [13]. Not only do these studies report temporal trends, but many also describe wide geographic variability in these measures and trends, which provides further evidence that environmental factors may play a role.

Flame-retardants are used in construction materials, furniture, plastics, electronics equipment, textiles, and other materials. Until recently, polybrominated diphenyl ethers (PBDEs) accounted for a large proportion of flame-retardants used in polyurethane foam and electronic applications [14]. However, in the past several years, common PBDE mixtures (i.e., pentaBDE and octaBDE) have been banned or voluntarily phased out in the United States and many parts of the world because of their persistence, bioaccumulation, and evidence for adverse health effects including endocrine disruption and altered fetal development [15, 16]. Thus, the use of alternate flame-retardants has been on the rise [17], as has scrutiny related to the potential environmental and human health consequences of alternate flame-retardants. Compared with PBDEs and other brominated flame-retardants (e.g., hexabromocyclododecane and tetrabromobisphenol A), organophosphorus (OP) flame-retardants have received little attention with regard to human exposure and potential health effects. Trichloroalkyl phosphates, such as tris(1, 3-dichloro-2-propyl) phosphate (TDCPP), and triaryl phosphates, such as triphenyl phosphate (TPP), continue to be used as flame-retardants and plasticizers in a wide variety of applications resulting in widespread environmental dispersion [18]. Production and use of OP flame-retardants has surpassed that of PBDEs in Europe [18], and annual production of both TDCPP and TPP in the United States has been estimated to be between 10 and 50 million pounds per year [19]. As with PBDEs, OPs such as TDCPP and TPP are used as additive flame-retardants that can be released into the surrounding environment over time [20]. Recent studies have reported that concentrations of OP flame-retardants measured in house dust are on the same order of magnitude as PBDEs, and for some OPs, such as TPP, concentrations greatly exceed those of PBDEs [17, 18]. Given that house dust is a primary source of exposure to PBDEs [21, 22] and that PBDEs can be detected in the blood of nearly all individuals in the general population [23], human exposures to OP flame-retardants are also likely to be widespread.

The presence of TDCPP was reported in a significant proportion of human seminal plasma samples nearly three decades ago [24], and high doses of OP flame-retardants have been associated with adverse reproductive, neurologic, and other systemic effects (e.g., altered thyroid and liver weights) in laboratory animals [14, 20, 25]. To our knowledge, no studies exploring associations between nonoccupational exposure to OP flame-retardants and human health end points have been conducted to date, although evidence exists for endocrine and reproductive effects in relation to other OP compounds. In the present study, we measured two OP flame-retardants (TDCPP and TPP) in house dust and assessed relationships with hormone levels and semen quality parameters among men recruited from an infertility clinic as part of an ongoing study of environmental influences on reproductive health.

We explored relationships of TDCPP and TPP concentrations in house dust with hormone levels and semen quality parameters.

We analyzed house dust from 50 men recruited through a U.S. infertility clinic for TDCPP and TPP. Relationships with reproductive and thyroid hormone levels, as well as semen quality parameters, were assessed using crude and multivariable linear regression.

TDCPP and TPP were detected in 96% and 98% of samples, respectively, with widely varying concentrations up to 1.8 mg/g. In models adjusted for age and body mass index, an interquartile range (IQR) increase in TDCPP was associated with a 3% [95% confidence interval (CI), −5% to −1%) decline in free thyroxine and a 17% (95% CI, 4–32%) increase in prolactin. There was a suggestive inverse association between TDCPP and free androgen index that became less evident in adjusted models. In the adjusted models, an IQR increase in TPP was associated with a 10% (95% CI, 2–19%) increase in prolactin and a 19% (95% CI, −30% to −5%) decrease in sperm concentration.

OP flame-retardants may be associated with altered hormone levels and decreased semen quality in men. More research on sources and levels of human exposure to OP flame-retardants and associated health outcomes are needed.

METHODS

Subject Recruitment

Our study was conducted on a subset of men participating in an ongoing study of environmental factors in reproductive health. Details of subject recruitment have been described previously [26]. Briefly, men between 18 and 54 years of age were recruited from the Vincent Memorial Andrology laboratory at Massachusetts General Hospital and invited to participate in the study. Applicable requirements involving human subjects were followed. Institutional review board approval was obtained from each participating institution, and all subjects signed an informed consent. Participants included men from couples who were infertile due to a male factor, a female factor, or a combination of both. Approximately 65% of eligible men agreed to participate. Exclusionary criteria included prior vasectomy or current use of exogenous hormones.

Dust Sample Collection and Analysis

The use of vacuum bags from the homes of study participants is a validated and cost-effective method of collecting household dust samples for a range of organic and inorganic chemicals in epidemiologic studies [26, 27]. We collected used household vacuum bags from men participating in the parent study between 2002 and 2007; existing vacuum bags were collected in the home by participants upon enrollment in the study. We analyzed vacuum bags from 50 participants, selected independently of fertility status or projected exposure levels, for OP flame-retardants.

Internal and surrogate standards used in this study were purchased from Wellington Laboratories (Guelph, Ontario, Canada), Cambridge Isotope Laboratories (Andover, MA), and Chiron (Trondheim, Norway). TDCPP was purchased from Chem Service (West Chester, PA), and TPP (99% pure) was purchased from Sigma-Aldrich (St. Louis, MO). All solvents used throughout this study were HPLC (high-performance liquid chromatography) grade. Approximately 0.3–0.5 g sieved (150-μm screen) dust was accurately weighed, spiked with 50–100 ng of two internal standards [4′fluoro-2,

3′, 4, 6-tetrabromodiphenyl ether (F-BDE 69) and 13C-labeled decabromodiphenyl ether (13C-BDE 209)], and extracted in stainless steel cells using pressurized fluid extraction (ASE 300, Dionex Inc., Sunnyvale, CA). Cells were extracted three times with 50:50 dichloromethane:hexane at a temperature of 100°C and at 1, 500 psi. Final extracts were reduced in volume to approximately 1.0 mL using an automated nitrogen evaporation system (Turbo Vap II; Caliper Life Sciences, Hopkinton, MA). Dust extracts were purified by elution through a glass column containing 4.0 g 6% deactivated alumina. All analytes were eluted with 50 mL of a 50:50 mixture of dichloromethane:hexane. The final extract was then reduced in volume to 0.5 mL, and 50 ng of the recovery standard, 2, 2′, 3, 4, 5, 5′-hexachloro[13C12]diphenyl ether (13C-CDE 141), was added prior to gas chromatography-mass spectroscopy (GC/MS) analysis.

We analyzed samples using GC/MS operated in either electron impact mode for TPP, or electron capture negative ionization mode for TDCPP. A 0.25-mm (i.d.) × 15 m fused silica capillary column coated with 5% phenyl methylpolysiloxane (0.25 μm film thickness) was used for separation of the analytes. Pressurized temperature vaporization injection was employed in the GC. The inlet was set to a temperature of 80°C for 0.3 min; a 600°C/min ramp to 275°C was then employed to efficiently transfer the samples to the head of the GC column. The oven temperature program was held at 40°C for 1 min followed by a temperature ramp of 18°C/min to 250°C, followed by a temperature ramp of 1.5°C/min to a temperature of 260°C, followed by a final temperature ramp of 25°C/min to 300°C, which was held for an additional 20 min. The transfer line temperature was maintained at 300°C, and the ion source was held at 200°C. 13C-BDE 209 was monitored by m/z 494.6 and 496.6; TDCPP was quantified by monitoring m/z 319 and 317, and TPP was quantified by monitoring m/z 326 and 325. As further confirmation, all ion ratios were monitored and were within 20% of their expected values compared with authentic standards.

As part of our quality assurance criteria, we examined levels of these specific analytes in laboratory blanks (n = 4), replicate samples (n = 3), and in matrix spikes (n = 3). Sample measurements were blank-corrected by subtracting the average level measured in the laboratory blanks. Blank levels for TDCPP and TPP were 11.7 ± 6.6 and 15.7 ± 11.9 ng, respectively. Method detection limits were calculated as three times the standard deviation of the blank levels. Matrix spikes were prepared by adding between 25 and 100 ng TDCPP and TPP to ASE cells filled with sodium sulfate powder. Matrix spikes were extracted using the same method used for dust and examined for percent recovery using 50 ng 13C-CDE 141 as an internal standard. Recoveries averaged 86 ± 7 and 89 ± 2% for TDCPP and TPP, respectively.

Serum Hormones

One nonfasting blood sample was drawn and centrifuged, and the serum was stored at −80°C until analysis. The hormone analytical methods have been described previously (Meeker et al. 2007). Briefly, we measured testosterone directly using the Coat-A-Count radioimmunoassay kit (Diagnostics Products, Los Angeles, CA); sex hormone binding globulin (SHBG) using a fully automated chemiluminescent immunometric assay (Immulite; DPC, Inc., Los Angeles, CA); and inhibin B using a commercially available, double antibody, enzyme-linked immunosorbent assay (Oxford Bioinnovation, Oxford, UK). Serum luteinizing hormone (LH), follicle-stimulating hormone (FSH), estradiol, prolactin, free thyroxine (T4), total triiodothyronine (T3), and thyrotropin (TSH) concentrations were determined by microparticle enzyme immunoassay using an automated Abbott AxSYM system (Abbott Laboratories, Chicago, IL). The free androgen index (FAI) was calculated as the molar ratio of total testosterone to SHBG.

Semen Quality

Semen samples were analyzed for sperm concentration and motion parameters by a computer-aided semen analyzer (HTM-IVOS, version 10HTM-IVOS; Hamilton-Thorne Research, Beverly, MA). Setting parameters and the definition of measured sperm motion parameters for the computer-aided semen analyzer were established by manufacturer. To measure both sperm concentration and motility, 5 μL of semen from each sample was placed into a prewarmed (37°C) Makler counting chamber (Sefi Medical Instruments, Haifa, Israel). We analyzed a minimum of 200 sperm cells from at least four different fields from each specimen. Motile sperm was defined as World Health Organization (WHO) grade a sperm (rapidly progressive with a velocity ≥ 25 μm/sec at 37°C) and grade b sperm (slow/ sluggish progressive with a velocity ≥ 5 μm/sec but < 25 μm/sec) (WHO 1999). For sperm morphology, at least two slides were made for each fresh semen sample. The resulting thin smear was allowed to air dry for 1 hr before staining with a Diff-Quik staining kit (Dade Behring AG, Dudingen, Switzerland). We performed morphologic assessment with a Nikon microscope using an oil immersion 100× objective (Nikon Company, Tokyo, Japan). We counted a minimum of 200 sperm cells from two slides for each specimen.

Statistical Analysis

We analyzed data using SAS, version 9.1 (SAS Institute Inc., Cary, NC). In preliminary analyses, Pearson correlation coefficients were calculated to assess bivariate relationships between OP concentrations in house dust, serum hormone levels, and semen quality parameters. Spearman rank correlation coefficients were also calculated and were compared with the Pearson correlations for consistency. Spearman correlations are presented only in instances where they were inconsistent with the Pearson correlation. Multivariable linear regression was then used to assess these relationships while adjusting for potential confounding variables. Serum levels of several hormones (testosterone, inhibin B, estradiol, free T4, and total T3), sperm motility, and sperm morphology closely approximated normality and were used in statistical models untransformed, whereas the distributions of several other hormones (FSH, LH, SHBG, FAI, prolactin, and TSH) and sperm concentration were skewed right and transformed to the natural log (ln) for statistical analyses. OP concentrations in dust were also ln-transformed. Dust samples with nondetectable OP concentrations were assigned a value equal to one-half the limit of detection (LOD). All multivariable models were adjusted for age and body mass index (BMI) as continuous variables. The models for semen quality parameters were also adjusted for abstinence period leading up to the collection of the semen sample, entered as a 5-level ordinal variable. Current smoking, time of day, season in which blood samples were collected from participants, and alcohol intake (number of drinks per week) were also considered but did not act as confounders and were not retained in the final models. To improve interpretability, the regression coefficients were exponentiated and expressed as a percent change in the dependent variable (i.e., change in hormone levels or semen quality parameter relative to the study population median) for an interquartile range (IQR) increase in dust OP concentration.

RESULTS

Of the 50 dust samples collected, we detected TDCPP and TPP in 48 (96%) and 49 (98%) samples, respectively. The distribution of measured TDCPP and TPP concentrations are presented in Table 1. Concentrations of TDCPP and TPP were positively (right)-skewed and varied widely, from < LOD (107 and 173 ng/g, respectively) to maximum values of 56 μg/g and 1, 800 μg/g, respectively. The distribution of age, BMI, serum hormone levels, and semen quality parameters among the men in the study are also presented in Table 1.

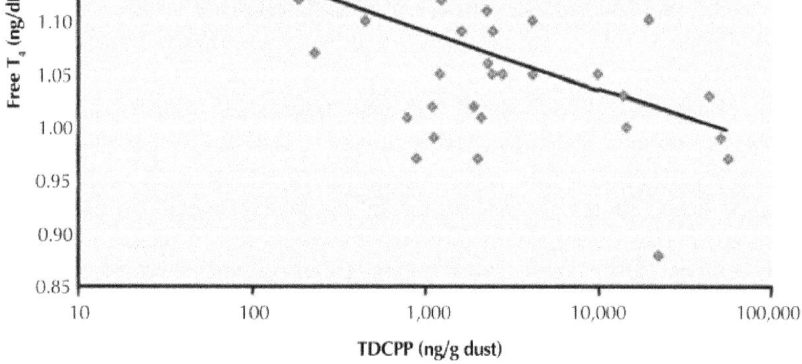

FIGURE 1 Scatterplot of TDCPP in house dust and serum free T_4 ($n = 38$; $r = -0.50$; $p = 0.002$). The inverse relationship remained when one outlying free T_4 value (0.88 ng/dL) was excluded from the analysis ($r = -0.45$, $p = 0.005$).

Semen quality parameters were available for all 50 men with OP dust measures, and hormone levels were available for 38 of the men. In preliminary correlation analysis, TDCPP and TPP were moderately correlated (Pearson r = 0.56, p = 0.0001; Spearman r = 0.33, p = 0.02). TDCPP was inversely associated with free T4 (Figure 1; correlation coefficients in the figures are Pearson correlations) and positively associated with prolactin (Figure 2). We also found a suggestive inverse association between TDCPP and FAI, as shown in Figure 3, although the Spearman rank correlation for this relationship was stronger than the Pearson correlation (Spearman r = −0.33, p = 0.04). TPP was positively correlated with prolactin (not shown; Pearson r = 0.37; p = 0.02) and inversely associated with sperm concentration (Figure 4). The inverse association remained when the three men with a sperm concentration below the commonly used reference criteria of 20 million sperm/mL [29] were removed from the analysis (Pearson r = −0.31, p = 0.03).

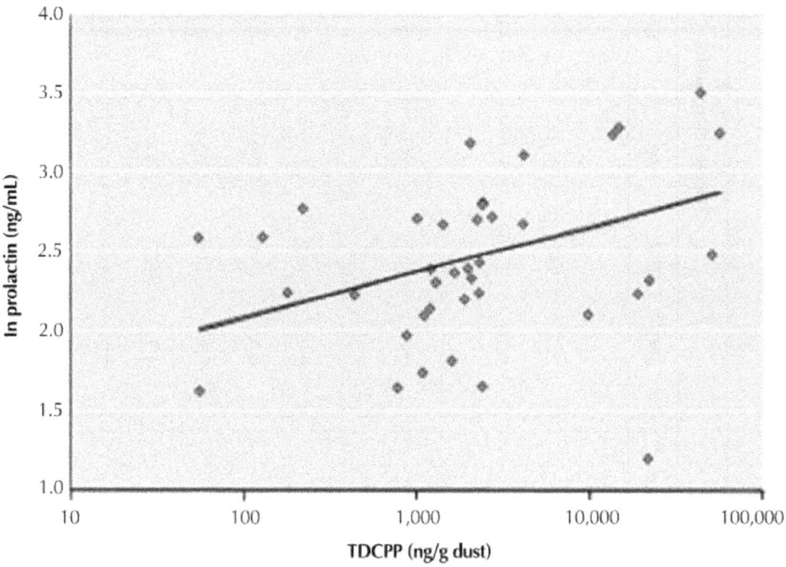

FIGURE 2 Scatterplot of TDCPP in house dust and serum prolactin (n = 38; r = 0.43; p = 0.0007).

In multivariable linear regression models adjusted for age and BMI, the inverse association between TDCPP and free T4 and the positive association between TDCPP and prolactin remained (Table 2). An IQR increase in dust TDCPP concentration was associated with a 2.8% [95% confidence interval (CI), −4.6% to −1.0%] decline in free T4 relative to the population median. This relationship was robust to the exclusion of an outlying free T4 value (0.88 ng/dL) from the model (not shown; p = 0.01). An IQR increase in TDCPP was also associated with a significant 17.3% (95% CI, 4.1–32.2%) increase in serum prolactin. The suggestive inverse association between TDCPP and FAI was no longer evident in the multivariate models (Table 2). The inconsistency between Pearson and Spearman correlations in the preliminary analysis led us to assess the potential for influential values in the results of parametric models. When excluding three subjects that had studentized residuals of > 2 or less than −2 in the original model, an IQR increase in TDCPP

TABLE 1 Distribution of TDCPP and TPP in House Dust, and Participant Age, BMI, Hormone Levels, and Semen Quality Parameters.

				Selected percentiles					
Variable	***n***	**Mean**	**Minimum**	**10th**	**25th**	**50th**	**75th**	**90th**	**Maximum**
OP									
TDCPP (ng/g dust)	50	1, 880a	< 107b	202	937	1, 752	2, 959	20, 415	56, 090
TPP (ng/g dust)	50	7, 400a	< 173b	760	3, 100	5, 470	9, 830	208, 200	1, 798, 100
Age (years)	50	36.5	28	31	33	37	40	42	46
BMI (kg/m^2)	50	26.8	20.3	21.6	23.0	25.9	29.7	33.0	42.4
Hormone									
Free T_4 (ng/dL)	38	1.07	0.88	0.97	1.02	1.07	1.13	1.19	1.23
Total T_3 (ng/mL)	38	0.92	0.74	0.76	0.80	0.93	1.02	1.10	1.23
TSH (μIU/mL)	38	1.43a	0.45	0.64	1.04	1.34	1.91	3.07	4.95
FSH (IU/L)	38	9.1a	3.1	4.3	6.5	8.8	13.1	18.0	29.5
LH (IU/L)	38	9.2a	3.7	5.2	6.6	9.7	11.7	16.1	20.4
Inhibin B (pg/mL)	38	209	67.6	115	142	177	241	308	702
Testosterone (ng/dL)	38	391	148	235	325	402	470	515	633

TABLE 1 *(Continued)*

				Selected percentiles					
Variable	***n***	**Mean**	**Minimum**	**10th**	**25th**	**50th**	**75th**	**90th**	**Maximum**
SHBG (nmol/mL)	38	24.4a	7.4	9.7	18.9	26.0	35.9	44.0	60.5
FAI	38	53.0a	19.5	32.6	42.1	49.9	69.4	92.9	143
Estradiol (pg/mL)	38	20.5	< 20.0b	< 20.0b	< 20.0b	21.5	28.0	38.0	48.0
Prolactin (ng/mL)	38	11.8a	5.1	5.7	9.2	11.0	15.2	25.9	33.6
Sperm concentration (10^6/mL)	50	70.5a	4.3	21.2	37.6	78.8	148	219	301
Sperm motility (% motile)	50	49.7	12	22	38	51	64	76	81
Sperm morphology (% normal)	50	6.9	0	2	4	7	10	14	17

[a]Geometric mean.
[b]LOD for TDCPP = 107 ng/g, for TPP = 173 ng/g, and for estradiol = 20 pg/mL.

was associated with a suggestive 6.3% decline in FAI (95% CI, −13.8 to 1.9%; p = 0.13).

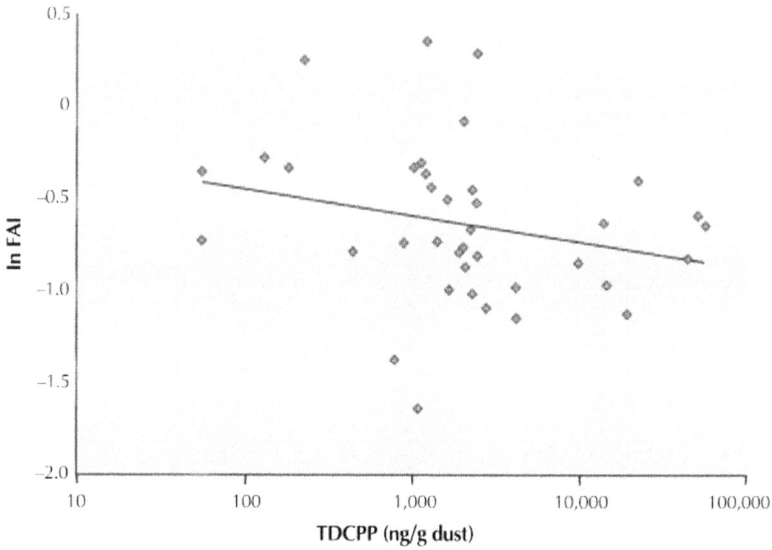

FIGURE 3 Scatterplot of TDCPP in house dust and serum FAI ($n = 38$; $r = 0.24$; $p = 0.14$).

TABLE 2 Adjusteda Regression Coefficients (95% CIs) for Percent Change in Hormone Level (Relative to Population Median) Associated with an IQR Increase in Dust TDCPP or TPP Concentration ($n = 38$).

	TDCPPb	p-Value	TPPb	p-Value
Free T_4	−2.8 (−4.6 to −1.0)	0.004	−1.1 (−2.4 to 0.2)	0.11
Total T_3	3.3 (−0.4 to 6.9)	0.08	2.0 (−0.5 to 4.5)	0.12
TSH[b]	3.1 (−11.8 to 20.5)	0.70	0.5 (−9.6 to 11.6)	0.93
FSH[b]	8.3 (−5.2 to 25.4)	0.23	4.7 (−4.5 to 14.9)	0.34
LH[b]	5.5 (−6.4 to 18.9)	0.36	3.5 (−3.4 to 12.2)	0.31
Inhibin B	−7.0 (−25.7 to 11.7)	0.45	−4.6 (−17.1 to 8.0)	0.46
Testosterone[c]	−2.1 (−8.5 to 4.3)	0.50	1.3 (−3.1 to 5.6)	0.55
Estradiol	8.7 (−5.3 to 22.7)	0.21	6.7 (−2.6 to 16.1)	0.15
SHBG[b]	2.7 (−8.9 to 17.3)	0.61	2.3 (−5.6 to 10.9)	0.53
FAI[b]	−5.2 (−15.9 to 5.5)	0.31	−1.2 (−8.8 to 5.9)	0.71
Prolactin[b]	17.3 (4.1 to 32.2)	0.008	9.7 (2.3 to 18.9)	0.02

[a]Adjusted for age and BMI.
[b]Variable ln-transformed in statistical analysis.
[c]Models for testosterone also adjusted for ln-transformed SHBG.

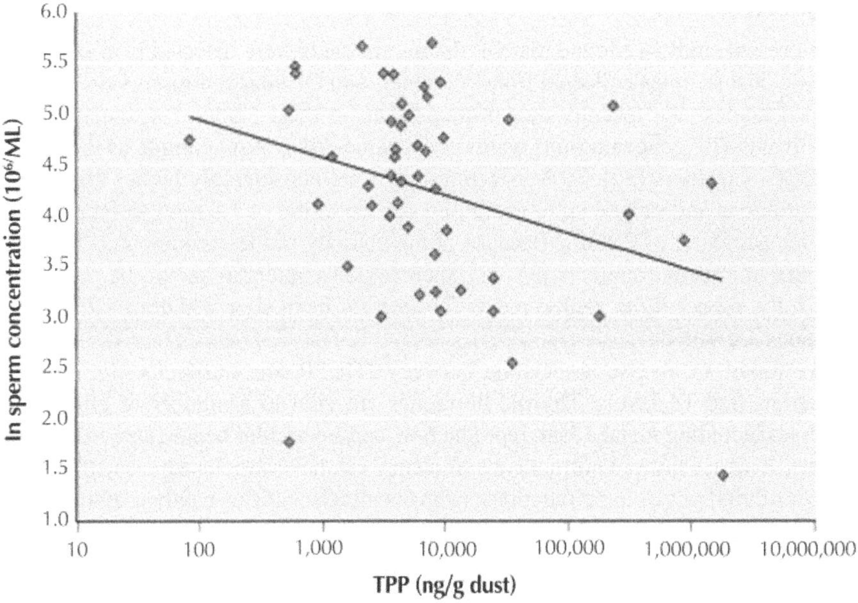

FIGURE 4 Scatterplot of TPP in house dust and sperm concentration (n = 50, r = 0.33, p = 0.02). The inverse association remained (r = −0.31, p = 0.03) when three men with sperm concentration of < 20 million/mL were excluded from the analysis.

The inverse association between TPP and sperm concentration was not sensitive to the adjustment for covariates (Table 3). When adjusting for age, BMI, and abstinence period, an IQR increase in TPP was associated with an 18.8% (95% CI, −30.1 to −4.5%) decline in sperm concentration. This inverse relationship remained when excluding three men with sperm concentration below the reference level of 20 million/mL (not shown; p = 0.02).

TABLE 3 Adjusteda Regression Coefficients (95% CIs) for Percent Change in Semen Quality Parameter (Relative to Population Median) Associated with an IQR Increase in Dust TDCPP or TPP Concentration (n = 50).

	TDCPPb	*p*-Value	TPPb	*p*-Value
Sperm concentration[b]	−13.6 (−32.0 to 8.3)	0.19	−18.8 (−30.1 to −4.5)	0.01
Sperm motility	−4.9 (−14.0 to 4.4)	0.29	−1.4 (−7.8 to 5.1)	0.67
Sperm morphology	−4.2 (−19.6 to 11.4)	0.59	−5.8 (−16.5 to 5.0)	0.28

[a]Adjusted for age, BMI, and abstinence period.
[b]Variable ln-transformed in statistical analysis.

DISCUSSION

In the present study, we found that OP flame-retardants were detected in nearly 100% of house dust samples collected from 50 homes, and OP concentrations varied widely between homes. We recently demonstrated that, in these samples, geometric mean TDCPP and TPP concentrations were on the same order of magnitude as the sum of 34 PBDE congeners [30]. TPP concentrations were considerably higher than PBDE concentrations and ranged up to maximum concentration of 1.8 mg/g compared with 0.04 mg/g for sum of PBDEs. Because of the ubiquity of these compounds in homes and other microenvironments [31–33], their toxicity potential should be considered more fully, especially as replacements for recently banned or withdrawn PBDE formulations are sought.

We found an inverse association between TDCPP concentrations in house dust and serum free T4 levels. Thyroid hormones are vital to a number of physiologic processes including metabolism, reproduction, cardiovascular health, and neurodevelopment. Because thyroid hormone insufficiency can have serious adverse effects on a number of vital physiologic functions, even chemicals that cause only a subtle shift in thyroid hormone levels should be considered carefully in terms of societal impact at the population level [34]. We also found positive relationships between both TDCPP and TPP with prolactin. Prolactin is a protein hormone that serves a number of important functions involving reproduction, metabolism, maintenance of homeostasis in immune responses, osmotic balance, and angiogenesis [35, 36] and is increasingly becoming used as a measure of neuroendocrine/dopaminergic function in environmental and occupational epidemiology studies [37, 38]. Because dopamine is responsible for inhibiting prolactin secretion, increased circulating levels of prolactin may reflect deficiencies in dopamine release, transport, or uptake [39]. Finally, in these data we also observed that an IQR increase in house dust TPP was associated with a substantial (19%) decline in sperm concentration. These findings may have substantial public health implications, given the likelihood of exposure to TPP among the general population, but more human and animal studies are needed to confirm our results.

Toxicology data relevant to the end points explored in the present study are limited [14, 20, 25]. TPP binds to the androgen receptor with moderate affinity [40] and activates enzymes involved in steroid hormone metabolism in vitro [41]. In addition, in rat studies butylated TPP has been associated with endocrine effects, including reduced male fertility and altered female reproductive cycles [42]. Rats dosed with high levels of TDCPP had altered thyroid weights and an increased thyroid/body weight ratio compared with control animals, and high-dose males demonstrated a significant increase in testicular interstitial-cell tumors and had increased histopathologic abnormalities in the testis, epididymis, and seminal vesicle [25].

We can also compare our findings with effects reported in association with other OP compounds. Consistent with our observation of an inverse association between TPP and sperm concentration, other OP flame-retardants and plasticizers such as tricresyl phosphate are reproductive toxicants that have caused marked reductions in male fertility and semen quality measures in laboratory animals [25]. Our findings are also consistent with several reports on the more well-studied OP insecticides. For example, in our previous work among a larger and overlapping group of men from the ongoing study, urinary levels of 3, 5, 6-trichloro-2-pyridinol, a urinary metabolite of the OP insecticides chlorpyrifos and chlorpyrifos-methyl, were associated with increased prolactin [43] and reduced free T4 [44], FAI [45], and sperm motility [46]. A variety of OP pesticides have been linked to decreased semen quality in studies of occupationally exposed men [47–50]. Case studies of patients with acute OP pesticide poisoning have also reported decreased levels of circulating thyroid hormones and increased prolactin [51, 52]. This is consistent with animal studies that reported inverse associations between chlorpyrifos and T4 levels in mice and sheep [53, 54], and studies that reported increased prolactin levels, decreased dopamine and testosterone levels, and damaged male reproductive organs in rats exposed to the OP insecticide quinalphos [55]. Thus, the relationships observed in the present study may be consistent with those reported for OP insecticides. However, compared with OP triesters used as insecticides, OP flame-retardants are more stable against hydrolysis, which makes them more persistent in the environment [18]. This potential for environmental persistence, combined with the widespread use and distribution of these compounds and the potential for adverse reproductive and neuroendocrine effects, serves as a further indication that more research is needed on the potential health effects resulting from exposure to OP flame-retardants.

Our study has a number of limitations. First, the sample size was relatively small, which may have limited our ability to detect subtle associations between OP levels and hormone or semen quality markers. Second, because this represents the first study to assess the relationship between OP flame-retardants and endocrine/reproductive outcomes in humans, a number of comparisons were made. Thus, the observation of a statistical relationship due to chance cannot be ruled out. Third, the study was cross-sectional in nature, so we cannot make conclusions regarding the temporality of the relationships observed. Fourth, we cannot rule out the presence of unmeasured confounders or coexposures that may explain our reported findings. However, we considered a number of potential confounding factors in the multivariable models. In addition, we previously reported that TDCPP and TPP were only moderately correlated with PBDE congeners and other brominated flame-retardants (all correlation coefficients were ≤ 0.4) [17]. Finally, we used house dust concentrations to estimate exposure to TDCPP and TPP. Because data on the prevalence, sources, and pathways of human exposure to TDCPP and TPP are lacking, this approach may be associated with exposure measurement error. However, if the measurement error were nondifferential, it would be expected to dilute exposure–outcome relationships toward the null [57]. Also, TDCPP and TPP are used (in a manner similar to that of PBDEs) as additive flame-retardants in furniture foams, textiles, plastics, and electronics, which may result in exposure sources and pathways similar to those of PBDEs in the home and in other microenvironments where house dust plays a primary role in aggregate exposure [21, 22]. Sensitive and specific biomarkers of exposure to these OP flame-retardants are needed, as are studies comparing the contribution of OP concentrations in house dust and other environmental media (e.g., air, diet) to biomarker levels to determine important pathways of exposure and the most relevant media and time windows for the estimation of exposure in epidemiologic studies.

CONCLUSION

We found evidence that concentrations of OP flame-retardants in house dust may be associated with altered hormone levels and decreased sperm concentration. More research is needed to determine the extent and sources of human exposure to OP flame-retardants and associated effects on human health.

KEYWORDS

- **Epidemiology**
- **Interquartile range**
- **Organophosphate**
- **Triphenyl phosphate**
- **Tris(1, 3-dichloro-2-propyl) phosphate**

REFERENCES

1. Diamanti-Kandarakis E, Bourguignon JP, Giudice LC, Hauser R, Prins GS, Soto AM, et al. Endocrine-disrupting chemicals: an Endocrine Society scientific statement. Endocrinology Review 2009; 30:293–342.
2. Carlsen E, Giwercman A, Keiding N, Skakkebaek NE. Evidence for decreasing quality of semen during past 50 years. BMJ 1992; 305:609–613.
3. Swan SH, Elkin EP, Fenster L. The question of declining sperm density revisited: an analysis of 101 studies published 1934–1996. Environmental Health Perspectives 2000; 108:961–966.
4. Andersson AM, Jensen TK, Juul A, Petersen JH, Jrgensen T, Skakkebaek NE. Secular decline in male testosterone and sex hormone binding globulin serum levels in Danish population surveys. Journal of Clinical Endocrinological Metabolism 2007; 92:4696–4705.
5. Travison TG, Araujo AB, O'Donnell AB, Kupelian V, McKinlay JB. A population-level decline in serum testosterone levels in American men. Journal of Clinical Endocrinological Metabolism 2007; 92:196–202.
6. Paulozzi LJ. International trends in rates of hypospadias and cryptorchidism. Environmental Health Perspectives 1999; 107:297–302.
7. Adami HO, Bergström R, Möhner M, Zatonski W, Storm H, Ekbom A, et al. Testicular cancer in nine northern European countries. International Journal of Cancer 1994; 59:33–38.
8. Bergström R, Adami HO, Möhner M, Zatonski W, Storm H, Ekbom A, et al. Increase in testicular cancer incidence in six European countries: a birth cohort phenomenon. Journal of the National Cancer Institute 1996; 88:727–733.
9. Huyghe E, Matsuda T, Thonneau P. Increasing incidence of testicular cancer worldwide: a review. Journal of Urolology 2003; 170:5–11.
10. Davies L, Welch HG. Increasing incidence of thyroid cancer in the United States, 1973–2002. JAMA 2006; 295:2164–2167.
11. Enewold L, Zhu K, Ron E, Marrogi AJ, Stojadinovic A, Peoples GE, et al. Rising thyroid cancer incidence in the United States by demographic and tumor characteristics, 1980–2005 Cancer Epidemiology Biomarkers Preview 2009; 18:784–791.
12. Harris KB, Pass KA. Increase in congenital hypothyroidism in New York State and in the United States. Molecular Genetic Metabolism 2007; 91:268–277.
13. Hertz-Picciotto I, Delwiche L. The rise in autism and the role of age at diagnosis. Epidemiology 2009; 20:84–90.

14. Babich JA. CPSC staff preliminary risk assessment of flame retardant (fr) chemicals in upholstered furniture foam. Bethesda, MD: Consumer Product Safety Commission; 2006, http://www.cpsc.gov/library/foia/foia07/brief/ufurn2.pdf [accessed 27 January 2010].
15. Meeker JD, Johnson PI, Camann D, Hauser R. Polybrominated diphenyl ether (PBDE) concentrations in house dust are related to hormone levels in men. Sci Total Environment 2009a; 407:3425–3429.
16. Meeker JD, Sathyanarayana S, Swan SH. Phthalates and other additives in plastics: human exposure and associated health outcomes. Philos Trans R Soc Lond B Biol Sci 2009b; 364:2097–2113.
17. Stapleton HM, Allen JG, Kelly SM, Konstantinov A, Klosterhaus S, Watkins D, et al. Alternate and new brominated flame-retardants detected in U.S. house dust. Environmental Science Technology 2008; 42:6910–6916.
18. Reemtsma T, Quintana JB, Rodil R, Garcia-Lopez M, Rodriguez I. Organophosphorus flame-retardants and plasticizers in water and air I. Occurrence and fate. Trends Anal Chem 2008; 27:727–737.
19. U.S. EPA. Inventory Update Reporting (IUR). IUR Data. Washington, DC: U.S. Environmental Protection Agency; 2006, http://www.epa.gov/oppt/iur/tools/data/ [accessed 8 July 2009].
20. U.S. EPA. Furniture flame retardancy partnership: environmental profiles of chemical flame-retardant alternative for low-density polyurethane foam. EPA 742-R-05-002. Washington, DC: U.S. Environmental Protection Agency; 2005, http://www.epa.gov/dfe/pubs/flameret/ffr-alt.htm [accessed 27 January 2010].
21. Johnson-Restrepo B, Kannan K. An assessment of sources and pathways of human exposure to polybrominated diphenyl ethers in the United States. Chemosphere 2009; 76:542–548.
22. Lorber M. Exposure of Americans to polybrominated diphenyl ethers. Journal of Expo Scientific Environmental Epidemiology 2008; 18:2–19.
23. Sjodin A, Wong LY, Jones RS, Park A, Zhang Y, Hodge C, et al. Serum concentrations of polybrominated diphenyl ethers (PBDEs) and polybrominated biphenyl (PBB) in the United States population: 2003–2004. Environmental Science Technology 2008; 42:1377–1384.
24. Hudec T, Thean J, Kuehl D, Dougherty RC. Tris(dichloropropyl) phosphate, a mutagenic flame retardant: frequent cocurrence in human seminal plasma. Science 1981; 211:951–952.
25. NRC (National Research Council). Toxicological risks of selected flame-retardant chemicals. Washington, DC: National Research Council, National Academies Press; 2000.
26. Meeker JD, Godfrey-Bailey L, Hauser R. Relationships between serum hormone levels and semen quality among men from an infertility clinic. Journal of Andrology 2007; 28:397–406.
27. Colt JS. Comparison of pesticides and other compounds in carpet dust samples collected from used vacuum cleaner bags and from a high-volume surface sampler. Environmental Health Perspectives 1998; 106:721–724.
28. Colt JS, Gunier RB, Metayer C, Nishioka MG, Bell EM, Reynolds P, et al. Household vacuum cleaners vs. the high-volume surface sampler for collection of carpet dust samples in epidemiologic studies of children. Environmental Health 2008; 7:6.
29. World Health Organization. WHO laboratory manual for examination of human semen and semen–cervical mucus interaction. Cambridge, UK: Cambridge University Press; 1999.
30. Stapleton HM, Klosterhaus S, Eagle S, Fuh J, Meeker JD, Blum A, et al. Detection of organophosphate flame-retardants in furniture foam and U.S. house dust. Environmental Science Technology 2009; 43:7490–7495.
31. Marklund A, Andersson B, Haglund P. Screening of organophosphorus compounds and their distribution in various indoor environments. Chemosphere 2003; 53:1137–1146.
32. Marklund A, Andersson B, Haglund P. Organophosphorus flame-retardants and plasticizers in air from various indoor environments. Journal of Environment Monitor 2005; 7:814–819.
33. Takigami H, Suzuki G, Hirai Y, Ishikawa Y, Sunami M, Sakai S. Flame-retardants in indoor dust and air of a hotel in Japan. Environ Int 2009; 35:688–693.

34. Miller MD, Crofton KM, Rice DC, Zoeller RT. Thyroid-disrupting chemicals: interpreting upstream biomarkers of adverse outcomes. Environmental Health Perspectives 2009; 117:1033–1041.
35. Ben-Jonathan N, LaPensee CR, LaPensee EW. What can we learn from rodents about prolactin in humans? Endocrinological Review 2008; 29:1–41.
36. Freeman ME, Kanyicska B, Lerant A, Nagy G. Prolactin: structure, function, and regulation of secretion. Physiology Review 2000; 80:1523–1631.
37. de Burbure C, Buchet JP, Leroyer A, Nisse C, Haguenoer JM, Mutti A, et al. Renal and neurologic effects of cadmium, lead, mercury, and arsenic in children: evidence of early effects and multiple interactions at environmental exposure levels. Environmental Health Perspectives 2006; 114:584–590.
38. Meeker JD, Rossano MG, Protas B, Diamond MP, Puscheck E, Daly D, et al. Multiple metals predict prolactin and thyrotropin (TSH) levels in men. Environmental Resources 2009c; 109:869–873.
39. Felt B, Jimenez E, Smith J, Calatroni A, Kaciroti N, Wheatcroft G, et al. Iron deficiency in infancy predicts altered serum prolactin response 10 years later. Pediatric Resources 2006; 60:513–517.
40. Fang H, Tong W, Branham WS, Moland CL, Dial SL, Hong H, et al. Study of 202 natural, synthetic, and environmental chemicals for binding to the androgen receptor. Chem Res Toxicology 2003; 16:1338–1358.
41. Honkakoski P, Palvimo JJ, Penttila L, Vepsalainen J, Auriola S. Effects of triaryl phosphates on mouse and human nuclear receptors. Biochemical Pharmacology 2004; 67:97–106.
42. Latendresse JR, Brooks CL, Flemming CD, Capen CC. Reproductive toxicity of butylated triphenyl phosphate and tricresyl phosphate fluids in F344 rats. Fundamentals of Applied Toxicology 1994; 22:392–399.
43. Meeker JD, Ravi SR, Barr DB, Hauser R. Circulating estradiol in men is inversely related to urinary metabolites of nonpersistent insecticides. Reproductive Toxicology 2008; 25:184–191.
44. Meeker JD, Barr DB, Hauser R. Thyroid hormones in relation to urinary metabolites of nonpersistent insecticides in men of reproductive age. Reproductive Toxicology 2006a; 22:437–442.
45. Meeker JD, Ryan L, Barr DB, Hauser R. Exposure to nonpersistent insecticides and male reproductive hormones. Epidemiology 2006b; 17:61–68.
46. Meeker JD, Ryan L, Barr DB, Herrick RF, Bennett DH, Bravo R, et al. The relationship of urinary metabolites of carbaryl/naphthalene and chlorpyrifos with human semen quality. Environmental Health Perspectives 2004; 112:1665–1670.
47. Padungtod C, Savitz DA, Overstreet JW, Christiani DC, Ryan LM, Xu X. Occupational pesticide exposure and semen quality among Chinese workers. Journal of Occupational Environmental Medicine 2000; 42:982–992.
48. Recio-Vega R, Ocampo-Gomez G, Borja-Aburto VH, Moran-Martinez J, Cebrian-Garcia ME. Organophosphorus pesticide exposure decreases sperm quality: association between sperm parameters and urinary pesticide levels. Journal of Applied Toxicology 2008; 28:674–680.
49. Yucra S, Gasco M, Rubio J, Gonzales GF. Semen quality in Peruvian pesticide applicators: association between urinary organophosphate metabolites and semen parameters. Environmenatl Health 2008; 7:59.
50. Yucra S, Rubio J, Gasco M, Gonzales C, Steenland K, Gonzales GF. Semen quality and reproductive sex hormone levels in Peruvian pesticide sprayers. International Journal of Occupational Environmental Health 2006; 12:355–361.
51. Guven M, Bayram F, Unluhizarci K, Kelestimur F. Endocrine changes in patients with acute organophosphate poisoning. Human Exp Toxicology 1999; 18:598–601.
52. Satar S, Satar D, Kirim S, Leventerler H. Effects of acute organophosphate poisoning on thyroid hormones in rats. Am J Ther 2005; 12:238–242.
53. De Angelis S, Tassinari R, Maranghi F, Eusepi A, Di Virgilio A, Chiarotti F, et al. Developmental exposure to chlorpyrifos induces alterations in thyroid and thyroid hormone levels without other toxicity signs in CD-1 mice. Toxicology Science 2009; 108:311–319.

54. Rawlings NC, Cook SJ, Waldbillig D. Effects of the pesticides carbofuran, chlorpyrifos, dimethoate, lindane, triallate, trifluralin, 2, 4-D, and pentachlorophenol on the metabolic endocrine and reproductive endocrine system in ewes. Journal of Toxicology and Environmental Health A 1998; 54:21–36.
55. Sarkar R, Mohanakumar KP, Chowdhury M. Effects of an organophosphate pesticide, quinalphos, on the hypothalamo-pituitary-gonadal axis in adult male rats. Journal of Reproductive Fertility 2000; 118:29–38.
56. Armstrong B. Exposure measurement error: consequences and design issues. In: Nieuwenhuijsen MJ, editor. Exposure assessment in occupational and environmental epidemiology. New York: Oxford University Press; 2004, pp. 181–200.

4 Bacterial and Fungal Microbial Biomarkers in House Dust

Joanne E. Sordillo, Udeni K. Alwis, Elaine Hoffman, Diane R. Gold, and Donald K. Milton

CONTENTS

INTRODUCTION

Measurement of fungal and bacterial biomarkers can be costly, but it is not clear whether home characteristics can be used as a proxy of these markers, particularly if the purpose is to differentiate specific classes of biologic exposures that have similar sources but may have different effects on allergic disease risk.

We evaluated home characteristics as predictors of multiple microbial biomarkers, with a focus on common and unique determinants and with attention to the extent of their explanatory ability.

In 376 Boston-area homes enrolled in a cohort study of home exposures and childhood asthma, we assessed the relationship between home characteristics gathered by questionnaire and measured gram-negative bacteria (GNB) (endotoxin and C10:0, C12:0, and C14:0 3-hydroxy fatty acids), gram-positive bacteria (GPB) (N-acetyl muramic acid), and fungal biomarkers [ergosterol and (1→6) branched, (1→3) β-d glucans] in bed and family room dust.

Home characteristics related to dampness were significant predictors of all microbial exposures; water damage or visible mold/mildew in the home was associated with a 20–66% increase in GNB levels. Report of cleaning the bedroom at least once a week was associated with reduced GNB, GPB, and fungi. Presence of dogs or cats predicted increases in home bacteria or fungi. The proportion of variance in microbial biomarkers explained by home characteristics ranged from 4.2% to 19.0%.

Despite their associations with multiple microbial flora, home characteristics only partially explain the variability in microbial biomarker levels and cannot substitute for specific microbial measurements in studies concerned with distinguishing effects of specific classes of microbes.

Evaluation of home characteristics by questionnaire is less costly than measurement of multiple microbial flora in the home. Home characteristics, particularly those associated with increased moisture, have been linked to respiratory symptoms [1–4]. Although it is known that increased dampness promotes the growth of microbial flora, the relationship of home characteristics to bacterial and fungal levels is imperfectly understood. Better understanding of that relationship can aid in evaluating the extent to which assessment of home characteristics can be a surrogate for measurement of bacteria and fungi in health effect studies, differentiating between types of microbial exposure, and identifying potentially modifiable conditions that may be the source of multiple exposures having protective or adverse health effects. Few studies have evaluated the relation of home characteristics to measures of multiple microbial exposures. Such an evaluation becomes increasingly important as we recognize that specific microbial agents from similar sources may differ in their effects. For example, we have shown that increased fungal exposures in the first year of life may increase the risk of allergic rhinitis [5], whereas endotoxin exposure may be a protective factor [6, 7].

Previously we demonstrated that sources of home dampness (humidifier use, water damage) predicted elevated levels of endotoxin, a gram-negative bacterial (GNB) biomarker, in the homes of infants enrolled in the Epidemiology of Home Allergens and Asthma Study [8]. In this follow-up study of these children at school age (mean age, 7 years), we assessed the relation of home characteristics (questionnaire data and allergen levels) and demographics to measures of multiple microbial exposures. We evaluated the explanatory power and positive and negative predictive values (PPV and NPV, respectively) of home characteristics in their associations with GNB, gram-positive bacteria (GPB), and fungi.

MATERIALS AND METHODS

Study Cohort

The Epidemiology of Home Allergens and Asthma Study is an ongoing longitudinal birth-cohort study of the effects of environmental exposures on the risk of allergy and asthma in children born to parents with histories of allergies and/or asthma [9]. The study was approved by the institutional review board of Brigham and Women's Hospital. We obtained written informed consent from the primary caregivers. Screening and recruitment of the families was conducted between September 1994 and June 1996. A detailed description of subject recruitment and early-life sampling for endotoxin and allergens in the home has been published previously [9, 10]. Briefly, families from metropolitan Boston, Massachusetts, were recruited at a major Boston hospital immediately after the birth of the index child. Exclusion criteria were gestational age < 36 weeks, congenital abnormalities, hospitalization in the neonatal intensive care unit, maternal age < 18 years, and a plan to move within the next year.

Home Visits and Microbial Biomarker Sample Collection

When the index children were of school age (mean age, 7 years), we conducted a home visit and collected three dust samples, two from the family room and one from the index child's bed. We used a Eureka Mighty-Mite vacuum cleaner (model 3621; Eureka Co., Bloomington, IN) modified to hold 19 × 90 mm cellulose extraction thimbles. For bed dust, all layers of the bedding were vacuumed for a total of 10 min. For family room dust samples, we vacuumed both a 1-m2 area of the family room floor for 2 min and an upholstered chair commonly used by the index child for 3 min. We repeated this procedure to collect a second family room dust sample. Of the 382 homes visited, 376 had microbial biomarker assessment in at least one room. Microbial biomarker levels were measured in 354 family room samples and in 299 beds. Family room dust samples were analyzed first for allergen levels and then for microbial biomarkers. If the first family room dust sample was consumed by allergen measurement (n = 49), the second dust sample was used for microbial biomarker assessment. In a subset of the cohort (n = 29), microbial biomarker measurements were performed on both the first and second family room samples. Paired t-tests indicated no significant difference between the two samples taken from the family room.

We administered a detailed questionnaire about home characteristics, including dampness-related variables (type of building, use of humidifier and dehumidifier, mold, water damage, central air conditioning), socioeconomic status (SES; income, race/ethnicity), carpeting, pests, frequency of cleaning (number of times the bedroom was cleaned per week), and number and sex of the children in the household. Exposure to pets (cats, dogs) and cockroaches was assessed by questionnaire and by allergen quantification in house dust.

Dust Samples

Within 24 hr after dust collection, we weighed and sieved the dust through a 425-μm mesh sieve. We reweighed the fine dust and made aliquots for allergen and microbial marker analysis. Allergen measurements were given first priority in bed and family room dust. Of the 382 homes visited, 354 (93%) of the homes had sufficient family room dust and 294 (77%) had sufficient bed dust for microbial biomarker assessment.

Microbial Biomarker Assays

We determined endotoxin bioactivity using both kinetic Limulus amebocyte lysate (LAL) and recombinant factor C (rFC) (Alwis and Milton 2006) assays with the kinetic Limulus assay with resistant parallel line estimation (KLARE) method as previously described (Milton et al. 1992). LAL and rFC reagents were obtained from Cambrex (Walkersville, MD), reference standard endotoxin from U.S. Pharmacopoeia, Inc. (Rockville, MD), and control standard endotoxin from Associates of Cape Cod (Woods Hole, MA). We assayed dust samples for 3-hydroxy fatty acids (3-OHFAs) as biomarkers of lipopolysaccharide using a method described elsewhere [11]. The 3-OHFAs with the highest correlation to endotoxin (C10:0, C12:0, and C14:0, or "midchain length") were summed and reported as picomoles per milligram of dust.

To measure peptidoglycan, we quantified the component N-acetyl muramic acid (2-acetamido-3-O-[(R)-1-carboxyethyl]-2-deoxy-d-glucose; hereafter muramic acid) by a previously described gas chromatography/mass spectrometry (GC/MS) method, using 13C-labeled cyanobacteria as an internal standard [12]. Muramic acid levels mainly represent GPB (the peptidoglycan layer is approximately 10–15 times thicker in GPB than in GNB cell walls). Results were reported as nanograms of muramic acid per milligram of dust.

For fungi, we measured the biomarkers (1→6) branched (1→3)β-d-glucan (hereafter β-d-glucan) and ergosterol. For β-d-glucan measurement we used an enzyme-linked immunosorbent assay [13] and for ergosterol we used a previously described GC/MS method [14, 15]. Results were reported as nanograms per milligram of dust.

Data Analysis

Microbial biomarkers showed right-skewed distributions. Therefore, we performed log10 transformation of those measurements to obtain symmetrical, approximately Gaussian distributions. We used log-transformed data for analyses of microbial biomarkers as continuous outcomes. For both bed and family room dust, these microbial biomarkers included ergosterol level (nanograms per milligram of dust), β-d-glucan (nanograms per milligram of dust), midchain (C10:0, C12:0, C14:0) 3-OHFAs (picomoles per milligram of dust), endotoxin (endotoxin units per milligram of dust, by both LAL and rFC), and muramic acid (nanograms per milligram of dust). We created binary predictor variables using the data from the home characteristics questionnaire. For bed and family room biomarker models these predictors were dampness-related characteristics (type of building, use of humidifier and dehumidifier, visible mold/mildew, water damage, central air conditioning), pets (current cat ownership, dog ownership), pests (cockroach and mouse sightings in the past year), and carpeting (a potential reservoir for many microbial organisms). To avoid entering two collinear predictors in models, we created a composite variable for water damage and visible mold/mildew (water/mold index), equal to 1 for those with either or both home characteristics and 0 for those with neither. For bed dust biomarker models specifically, we also examined the influence of more than four stuffed animals on the bed, cleaning the bedroom once or more per week, and child's sex on microbial biomarker levels in mattress dust. Additionally, variables for a continuously burning pilot light and the total number of household children (total boys and total girls) were considered as predictors of family room biomarker levels.

To build multiple regression models for bed and family room microbial biomarkers, we first constructed reduced models containing a priori home characteristics (those identified from the literature) and those statistically significant ($p < 0.05$) in univariate models. Home characteristics that were significant additions to the reduced model ($p < 0.05$) were entered into the multiple regression model. After selection of home characteristics, we adjusted final models for seasonal effects. Multiple regression models incorporating allergen levels (≥ 20 μg/g Can f 1, ≥ 8 μg/g Fel d 1, detectable Bla g 1) in the place of questionnaire data (current dog ownership, cat ownership, cockroach sightings in the past year) are also shown.

In secondary analyses, predictive values of individual home characteristics as indicators of GNB, GPB, and fungi were also assessed for bed exposure. To perform these calculations, microbial biomarker levels were classified as either above or below the median. A priori we chose the median as the cut-point because we have demonstrated that microbial exposures above this threshold have been associated with respiratory health [7, 10]. We computed PPVs and NPVs to express the probability of being above or below the median level of microbial exposure, given the presence or absence of a particular home characteristic. [PPV is calculated by determining the percentage of homes with an elevated microbial biomarker level (i.e., > median ergosterol) among all homes with a given home characteristic (i.e., "dog ownership" or > 20 μg/g Can f 1). NPV is calculated by determining the percentage of homes with decreased microbial biomarker levels (i.e., < median ergosterol) among all homes lacking a given home characteristic (i.e., < 20 μg/g Can f 1)]. In predictive value calculations for GNB, elevated exposure was defined as greater than the median level of endotoxin (LAL) or midchain 3-OHFAs in bed dust. To evaluate whether the cut-point influenced the overall inference regarding the PPV and NPV of the home characteristics for microbial markers, we also examined a second cut-point at the 75th percentile.

RESULTS

Characteristics and Correlations

Only a small percentage of the population reported living in an apartment, earning < $35, 000/year, or using a dehumidifier or humidifier in the family room (Table 1). Other home characteristics such as dog or cat ownership, central air conditioning, mouse sightings, visible mold/mildew and/or water damage, and frequent cleaning were more prevalent (21.5–54.4%). Biomarker levels were higher in family room dust samples than in bed dust samples (Table 2). Correlations between midchain 3-OHFAs and endotoxin were moderate in both bed and family room dust samples (Table 3) (Pearson $r = 0.39$–0.45). Midchain 3-OHFAs were moderately correlated with β-d-glucan and ergosterol in the bed samples ($r = 0.42$ and 0.43) but were not as highly correlated in family room dust ($r = 0.23$ and 0.28). Muramic acid was most highly correlated with midchain 3-OHFAs. The fungal biomarkers ergosterol and β-d-glucan showed only a modest correlation in bed ($r = 0.32$) and family room ($r = 0.22$) samples.

TABLE 1 Characteristics of Homes with Bed and Family Room Dust Samples [n (%)].

Category	Characteristic	Homes with family room sample ($n = 354$)	Homes with bed dust sample ($n = 294$)
Dampness	Report of water damage in home or visible mold/mildew in room sampled (water/mold index)	112 (31.7)	88 (29.9)
	Humidifier use in room sampled	16 (4.5)	89 (30.4)
	Dehumidifier use in room sampled	13 (3.7)	1 (0.3)
	Central air	103 (29.2)	78 (26.5)
	Living in an apartment	16 (4.5)	14 (4.8)
Pets	Currently owns a dog	76 (21.5)	70 (23.8)
	Currently owns a cat	76 (21.5)	67 (22.8)
Pests	Any mouse sightings in past year	111 (31.3)	103 (35.0)
	Any cockroach sightings in the past year	10 (2.8)	3 (1.0)
Cleaning	Cleaning frequency ≥ once/week	NA	160 (54.4)
Reservoir for microbes	> Four stuffed animals on bed	NA	50 (17.0)
	Wall-to-wall carpet	113 (31.9)	108 (36.7)

TABLE 1 *(Continued)*

Category	Characteristic	Homes with family room sample ($n = 354$)	Homes with bed dust sample ($n = 294$)
SES	Income < $35, 000/year	12 (3.4)	11 (3.7)
	Black race/ethnicity	26 (7.4)	26 (8.9)
Other	Continuously burning pilot light	48 (13.6)	NA

NA, not analyzed.

TABLE 2 Biomarker Levels in Dust.

Category	Characteristic	*n*	Mean	Median	Interquartile range
GNB	Endotoxin (LAL) (EU/mg)				
	Bed dust	294	23.7	18.9	12.3–31.3
	Family room dust	354	64.3	38.6	25.6–58.9
	Endotoxin (rFC) (EU/mg)				
	Bed dust	294	8.8	5.9	3.2–10.6
	Family room dust	354	28.7	10.3	5.3–23.9
	Midchain 3-OHFAs (pmol/mg)				
	Bed dust	294	39.6	35.8	28.0–47.2
	Family room dust	354	64.6	60.3	48.9–75.6
GPB	Muramic acid (ng/mg)				
	Bed dust	293	69.4	62.6	45.0–79.4
	Family room dust	350	77.4	72.2	54.6–95.3
Fungi	β-d-Glucan (ng/mg)				
	Bed dust	294	19.4	16.8	11.6–24.6
	Family room dust	353	29.9	25.0	17.0–36.3
	Ergosterol (ng/mg)				
	Bed dust	294	1.3	1.0	0.6–1.7
	Family room dust	344	4.1	2.5	1.7–4.1

TABLE 3 Biomarker Correlations: Pearson Correlation (log10 biomarker level).

Category	Characteristic	β-d-Glucan	Ergosterol	Muramic acid	Endotoxin (rFC)	Endotoxin (LAL)	Midchain 3-OHFAs
Bed dust							
Fungi	β-d-Glucan	1.0					
	Ergosterol	0.32	1.0				
GPB	Muramic acid	0.18	0.24	1.0			
GNB	Endotoxin (rFC)	0.23	0.35	0.20	1.0		
	Endotoxin (LAL)	0.25	0.41	0.32	0.67	1.0	
	Midchain 3-OHFAs	0.42	0.43	0.26	0.39	0.45	1.0
Family room dust							
Fungi	β-d-Glucan	1.0					
	Ergosterol	0.22	1.0				
GPB	Muramic acid	0.15	0.13	1.0			
GNB	Endotoxin (rFC)	0.22	0.20	0.25	1.0		
	Endotoxin (LAL)	0.22	0.24	0.29	0.82	1.0	
	Midchain 3-OHFAs	0.23	0.28	0.35	0.42	0.39	1.0

[a]$p < 0.0001$ for all $r \geq 0.23$; $p < 0.05$ for all $r \geq 0.10$ and ≤ 0.22.

Categories of Predictors

Predictors fell into broad categories: dampness, pets, socioeconomic factors, cleaning, and number/sex of children in the household. For multiple regression models, Tables 4 and 5 show the percent change in microbial biomarker levels associated with these predictors. Multiple regression models explained a low to moderate portion of the variability in microbial biomarker levels (R^2 = 4.2–19.0%).

Characteristics related to dampness were the most consistent predictors of microbial biomarkers. The water/mold index predicted increased GNB biomarkers in both bed and family room dust. This variable was also associated with higher levels of the fungal biomarker ergosterol found in family room dust. Central air conditioning, a possible indicator of drier home conditions, predicted reductions in levels of GNB (endotoxin and midchain 3-OHFAs), GPB (muramic acid), and fungi (ergosterol and β-d-glucan). Living in an apartment, also associated with drier conditions, was a predictor of decreased fungal biomarkers in family room and bed dust.

Cleaning the bedroom at least once a week showed associations with lower levels of GNB (LAL endotoxin and midchain 3-OHFAs), GPB (muramic acid), and fungi (ergosterol). We also observed a similar trend for cleaning for endotoxin by rFC, but it did not achieve statistical significance in multiple regression models.

Although sources of dampness in both the bedroom and the family room predicted increases in all of the microbial biomarkers, the associations of pets with these biomarkers were less consistent. Regardless of whether ascertained by questionnaire or allergen level measurement, pets or pests had the same direction of association with levels of measured microbial biomarkers. However, in the bedroom, compared with reports of owning pets, high levels of dog and cat allergen were more strongly predictive of elevated GNB biomarkers. The direction of the association of dog with markers of mold differed by room, with dog predicting higher β-d-glucan levels in the bedroom but lower ergosterol and β-d-glucan levels in the family room. Black race/ethnicity was consistently associated with lower levels of microbial biomarkers, a predictor of lower GNB (rFC endotoxin) and fungi (β-d-glucan). Low income was associated with lower ergosterol levels but was not associated with other microbial biomarker levels.

For family room exposures, boys were associated with moderate increases in GNB, whereas girls were weakly associated with lower levels of GPB and fungi (β-d-glucan). However, the index child's sex was not a significant predictor of microbial biomarkers in individual bed dust samples.

Predictive Values for Home Characteristics

We employed home characteristics from the strongest categories of predictors as tests for high (> median) microbial biomarker levels. Elevated pet allergens, home dampness, and infrequent cleaning were associated with a 71–75% probability (PPV) of increased (> median) GNB. PPVs observed for pet allergen levels as predictors of elevated GPB and fungi were substantially lower (PPV, 54–66%). NPVs, or the probability of microbial exposure at less than the median threshold given the absence of a home characteristic, ranged from 36% to 54% for all microbial biomarker levels. Cut-point strongly influenced the predictive values of home characteristics. A cut-point at greater than the 75th percentile increased the NPV of dampness, pet, and cleaning characteristics to ≥ 70%.

TABLE 4 Bed Dust: percent difference in microbial biomarkers associated with predictors (home characteristics, demographics, and allergen levels) in multiple regression models (95% confidence interval for percent difference).

	GNB			GPB	Fungi	
Characteristic and model	**Recombinant factor C**	**LAL**	**Midchain 3-OH-FAs**	**Muramic acid**	**Ergosterol**	**β-d-Glucan**
Multiple regression models with questionnaire and allergen (pest/pet) predictors						
Dampness						
Water/mold	38.6 (11.1 to 72.9)*	22.9 (4.9 to 43.9)*	—	7.5 (−5.1 to 21.8)	17.7 (−1.1 to 40.1)	6.5 (−6.8 to 21.8)
Central air	—	—	−11.9 (−19.9 to −3.1)*	−11.4 (−22.0 to 0.7)	−29.3 (−41.1 to −14.7)*	−14.2 (−25.8 to −1.6)*
Apartment	—	—	—	—	—	−31.1 (−48.3 to −8.2)*
Pets						
Can f 1 ≥ 20 μg/g	16.1 (−9.4 to 48.7)	11.3 (−6.7 to 32.7)	13.7 (2.5 to 26.2)*	13.4 (−1.4 to 30.4)	—	20.6 (3.7 to 40.2)*
Fel d 1 ≥ 8 μg/g	36.9 (8.3 to 73.0)*	34.2 (13.5 to 58.5)*	—	—	—	—
Pests						
Bla g 1 detectable	—	—	—	—	−42.9 (−63.2 to −11.6)*	—

TABLE 4 *(Continued)*

	GNB			GPB	Fungi	
Characteristic and model	**Recombinant factor C**	**LAL**	**Midchain 3-OH-FAs**	**Muramic acid**	**Ergosterol**	**β-d-Glucan**
Cleaning						
≥ 1/week	−16.6 (−31.9 to 2.1)	−15.0 (−26.4 to −1.8)*	−9.3 (−17.1 to −1.8)*	−11.9 (−21.4 to −1.4)*	−15.1 (−27.6 to −0.4)*	—
Reservoir for microbes						
Wall-to-wall carpet	—	—	—	—	23.7 (4.0 to 47.1)*	—
> Four stuffed animals	—	—	13.2 (1.1 to 26.7)*	—	—	—
SES						
Black race/ethnicityb	−41.8 (−59.2 to −16.9)*	−18.6 (−36.8 to 5.0)	—	—	—	—
Seasonc						
Spring	20.6 (−8.5 to 32.9)	10.0 (−9.6 to 34.0)	−3.5 (−14.2 to 8.5)	21.5 (4.2 to 41.6)*	−15.1 (−31.5 to 5.2)	10.3 (−6.4 to 30.0)
Summer	3.3 (−19.7 to 32.9)	9.3 (−8.6 to 30.8)	34.5 (20.7 to 49.8)*	0.3 (−13.0 to 15.7)	8.6 (−10.9 to 32.3)	47.9 (27.1 to 72.0)*
Autumn	9.5 (−22.0 to 53.6)	3.8 (−18.4 to 32.2)	−2.2 (−15.2 to 12.9)	23.7 (2.0 to 49.9)*	−5.5 (−27.3 to 22.9)	16.6 (−4.8 to 42.7)

TABLE 4 *(Continued)*

	GNB			GPB	Fungi	
Characteristic and model	**Recombinant factor C**	**LAL**	**Midchain 3-OH-FAs**	**Muramic acid**	**Ergosterol**	**β-d-Glucan**
Model adjusted R^2	7.7	7.1	19.0	5.6	9.3	12.7
Multiple regression models with questionnaire predictors						
Dampness						
Water/mold	35.1 (8.0 to 68.9)*	20.3 (2.5 to 41.3)*	—	6.4 (−5.2 to 21.9)	17.2 (−1.7 to 39.8)	6.3 (−7.2 to 21.6)
Central air	—	—	−12.2 (−20.3 to −3.4)*	−10.9 (−22.2 to 0.4)	−30.4 (−42.4 to −15.9)*	−14.5 (−25.5 to −1.9)*
Apartment	—	—	—	—	—	−29.9 (−47.4 to −6.4)*
Pets						
Dog ownership	15.6 (−9.1 to 46.9)	9.1 (−8.1 to 29.6)	8.3 (−2.0 to 19.6)	11.8 (−3.7 to 25.9)	—	15.8 (0.2 to 33.9)*
Cat ownership	18.7 (−6.7 to 51.1)	17.2 (−1.4 to 39.3)	—	—	—	—
Pests						
Cockroach sighting	—	—	—	—	1.7 (−54.0 to 125.3)	—
Cleaning						
≥ 1/week	−17.2 (−32.5 to 1.5)	−15.7 (−27.1 to −2.4)*	−9.1 (−16.6 to −1.0)*	−11.4 (−21.2 to −1.1)*	−15.3 (−27.9 to −0.5)*	—

TABLE 4 *(Continued)*

	GNB			GPB	Fungi	
Characteristic and model	**Recombinant factor C**	**LAL**	**Midchain 3-OH-FAs**	**Muramic acid**	**Ergosterol**	**β-d-Glucan**
Reservoir for microbes						
Wall-to-wall carpet	—	—	—	—	20.7 (1.4 to 43.7)*	—
> Four stuffed animals	—	—	12.7 (0.7 to 26.3)*	—	—	—
SES						
Black race/ethnicity[b]	−41.5 (−59.1 to −16.2)*	−18.3 (−36.9 to 5.7)	—	—	—	—
Season[c]						
Spring	16.1 (−11.9 to 53.1)	5.9 (−13.1 to 29.1)	−2.5 (−13.1 to 9.3)	19.3 (2.2 to 39.2)*	−16.4 (−32.7 to 3.9)	10.8 (−6.1 to 30.6)
Summer	0.7 (−21.8 to 29.6)	6.7 (−11.0 to 27.9)	34.3 (20.8 to 49.3)*	−3.8 (−16.5 to 10.8)	8.0 (−11.6 to 31.9)	48.3 (27.4 to 72.6)*
Autumn	5.1 (−25.2 to 47.6)	−0.05 (−21.7 to 27.6)	−1.8 (−14.7 to 13.1)	19.8 (−0.7 to 44.6)	−7.1 (−28.8 to 21.2)	16.0 (−5.3 to 42.1)
Model adjusted R^2	6.2	4.2	18.0	7.8	9.3	12.7

[a]Absence of the home characteristic served as the reference category for each predictor unless otherwise noted. Dashes indicate that the characteristic was not entered into the model.
[b]Reference category is all other race/ethnicity classifications (white, Hispanic, Asian, and other).
[c]Reference category is winter.
*$p < 0.05$.

TABLE 5 Family Room Dust: percent difference in microbial biomarkers associated with predictors (home characteristics, demographics, and allergen levels) in multiple regression models (95% confidence interval for percent difference).

Characteristic and model	GNB			GPB	Fungi	
	Recombinant factor C	LAL	Midchain 3–OH-FAs	Muramic acid	Ergosterol	β-d-Glucan
Multiple regression models with questionnaire predictors						
Dampness						
Water/mold	65.6 (23.8 to 113.5)*	43.8 (20.5 to 71.5)*	—	5.6 (−4.4 to 16.7)	18.6 (−0.1 to 40.8)	9.4 (−5.2 to 26.2)
Central air	−28.2 (−44.7 to −3.4)*	−16.2 (−30.1 to 0.4)	—	−14.2 (−22.5 to −4.8)*	—	—
Apartment	10.8 (−41.2 to 107.9)	−15.0 (−43.5 to 27.9)	−15.1 (−30.5 to 3.7)	—	−39.2 (−58.5 to −11.0)*	−31.1 (−50.6 to −5.7)*
Pets						
Can f 1 ≥ 20 μg/g	21.7 (−15.5 to 56.1)	−0.9 (−18.7 to 21.0)	−2.4 (−5.2 to 18.4)	5.9 (−5.2 to 18.4)	−22.1 (−35.7 to −5.7)*	−12.1 (−25.1 to 3.1)
Fel d 1 ≥ 8 μg/g	—	—	−8.5 (−16.7 to 0.5)	—	—	—
Pests						
Bla g 1 detectable	—	—	—	—	—	−30.5 (−52.6 to 1.6)
Mouse sighting	—	—	—	−10.8 (−19.3 to −1.4)*	—	—

TABLE 5 *(Continued)*

Characteristic and model	GNB			GPB	Fungi	
	Recombinant factor C	LAL	Midchain 3–OH-FAs	Muramic acid	Ergosterol	β-d-Glucan
SES						
Black race/ethnicityb	−43.5 (−66.7 to −8.5)*	−6.4 (−32.5 to 29.8)	−12.1 (−25.1 to 3.2)	—	—	−22.9 (−40.5 to −0.2)*
Income < $35, 000/year	83.5 (−11.2 to 287.8)	27.0 (−21.2 to 104.7)	3.1 (−11.6 to 20.2)	—	−27.4 (−45.9 to −2.5)*	—
Number/sex of children						
Per boy	15.9 (0.5 to 33.7)*	12.5 (2.5 to 23.4)*	—	—	—	—
Per girl	—	—	—	−5.9 (−10.7 to −0.8)*	—	−8.2 (−14.8 to −0.9)*
Wall-to-wall carpet	—	−8.1 (−15.8 to 0.1)	—	—	—	—
Seasonc						
Spring	−8.3 (−34.9 to 29.0)	−6.7 (−25.2 to 16.4)	9.3 (−2.0 to 22.0)	3.8 (−8.2 to 17.4)	0.9 (−18.5 to 24.8)	1.2 (−15.3 to 20.9)
Summer	−13.3 (−37.2 to 19.8)	−10.3 (−27.2 to 10.6)	21.8 (9.7 to 35.3)*	−10.6 (−20.5 to 0.5)	16.5 (−5.1 to 43.1)	24.1 (4.8 to 46.8)*
Autumn	1.1 (−33.0 to 52.7)	−15.2 (−35.1 to 10.7)	−1.9 (−14.0 to 12.0)	−0.6 (−14.4 to 15.4)	13.2 (−12.3 to 46.0)	16.5 (−5.9 to 44.2)

TABLE 5 (Continued)

Characteristic and model	GNB			GPB	Fungi	
	Recombinant factor C	LAL	Midchain 3–OH-FAs	Muramic acid	Ergosterol	β-d-Glucan
Model adjusted R^2	8.3	4.9	6.9	4.9	4.6	6.0
Multiple regression models with questionnaire and allergen (pest/pet) predictors						
Dampness						
Water/mold	61.3 (22.8 to 111.7)*	42.5 (19.4 to 70.0)*	—	5.1 (−4.8 to 16.1)	18.0 (−0.7 to 40.3)	12.5 (−2.4 to 29.7)
Central air	−26.4 (−44.3 to −2.7)*	−15.7 (−29.7 to 1.0)	—	−13.8 (−22.2 to −4.4)*	—	—
Apartment	10.0 (−41.4 to 106.4)	−14.7 (−43.3 to 28.4)	−15.3 (−30.6 to 3.4)	—	−38.1 (−57.5 to −9.3)*	−29.1 (−48.6 to −2.3)*
Pets						
Dog ownership	23.5 (−9.7 to 68.7)	7.2 (−12.5 to 31.3)	−8.0 (−16.7 to 1.5)	10.9 (−1.0 to 24.2)	−19.1 (−33.5 to −1.6)*	−15.1 (−27.9 to 0.1)
Cat ownership	—	—	−8.2 (−16.9 to 1.4)	—	—	—
Pests						
Cockroach sighting	—	—	—	—	—	−38.8 (−58.9 to −8.9)*

TABLE 5 *(Continued)*

Characteristic and model	GNB			GPB	Fungi	
	Recombinant factor C	**LAL**	**Midchain 3–OH-FAs**	**Muramic acid**	**Ergosterol**	**β-d-Glucan**
Mouse sighting	—	—	—	−10.8 (−19.3 to −1.4)*	—	—
SES						
Black race/ethnicityb	−44.8 (−66.7 to −8.7)*	−5.9 (−32.1 to 30.4)	−11.6 (−24.7 to 3.7)	—	—	−24.0 (−41.2 to −1.9)*
Income < $35, 000/year	85.4 (−11.2 to 287.0)	26.3 (−21.5 to 103.8)	2.6 (−12.0 to 19.6)	—	−27.4 (−46.0 to −2.4)*	—
Number/sex of children						
Per boy	16.3 (0.8 to 34.1)*	13.0 (3.1 to 24.0)*	—	—	—	—
Per girl	—	—	—	−6.0 (−10.8 to −0.9)*	—	−8.7 (−15.3 to −1.5)*
Wall-to-wall carpet	—	−7.9 (−15.6 to 0.5)	—	—	—	—
Seasonc						
Spring	1.5 (−32.7 to 53.1)	−6.4 (−25.0 to 16.9)	9.0 (−2.3 to 21.6)	4.5 (−7.6 to 18.1)	−0.3 (−19.5 to 23.5)	1.2 (−17.2 to 18.1)
Summer	−13.8 (−37.6 to 19.0)	−10.3 (−27.6 to 10.1)	22.6 (10.4 to 36.2)*	−10.9 (−20.7 to 0.2)	16.6 (−5.2 to 43.3)	22.5 (3.5 to 44.9)*

TABLE 5 *(Continued)*

Characteristic and model	GNB			GPB	Fungi	
	Recombinant factor C	LAL	Midchain 3–OH-FAs	Muramic acid	Ergosterol	β-d-Glucan
Autumn	−7.3 (−34.0 to 30.5)	−15.0 (−34.9 to 11.0)	−1.5 (−13.7 to 12.3)	−0.3 (−14.1 to 15.7)	13.4 (−12.2 to 46.4)	13.8 (−8.0 to 40.8)
Model adjusted R^2	5.8	7.8	7.5	5.5	6.5	6.9

[a]Absence of the home characteristic served as the reference category for each predictor unless otherwise noted. Dashes indicate that the characteristic was not entered into the model.
[b]Reference category is all other race/ethnicity classifications (white, Hispanic, Asian, and other).
[c]Reference category is winter.
*$p < 0.05$.

DISCUSSION

The association between home characteristics (particularly dampness) and respiratory health is thought to be mediated by exposures to bacteria and fungi. Although many studies have examined the link between home environment and individual microbial biomarkers (i.e., endotoxin), this work simultaneously assessed the associations between home characteristics and the microbial milieu (GNB, GPB, and fungi). Our findings suggest the following: Presence of damp environment, pets, and less frequent cleaning partially explain higher levels of home bacteria and mold; sources of microbial exposures (represented by home characteristics) overlap for GNB, GPB, and fungi and cannot be used to distinguish between these groups of organisms.

Categories of Predictors

Four categories of predictors [dampness, pets, cleaning, and demographics (SES and sex)] were consistently associated with the home microbial environment. For GNB, GPB, and fungi, indoor home characteristics associated with moisture were the strongest predictors in multiple regression models. The link between dampness and GNB has been shown in previous studies, where water damage and high humidity conferred elevated levels of endotoxin or 3-OHFAs [16–18]. The inverse relationship between GNB levels and central air conditioning, a potential indicator of dryness, reported in the present work supports a similar published association [19]. Numerous studies have also identified moisture as a promoter of increased fungal levels in the indoor environment [1, 20]. The causes and consequences of dampness itself are heterogeneous in nature, and moisture damage may involve a variety of different building materials. Although dampness has been shown to increase GNB, GPB, and fungi, the characteristics of microbial colonization (e.g., bacteria or fungi, taxa, genus, species) resulting from damp conditions will depend on the nature of the individual moisture problem [21]. Although indicators of home dampness may relate to total microbial burden, they cannot provide insight into the specific type of microbial growth.

Pet ownership and pet allergen levels were positively associated with higher levels of fungi (β-d-glucan) and GNB (midchain 3-OHFAs and endotoxin). The importance of pets as contributors to the microbial environment has been documented in other studies [19, 22–25]. In the present work, dog exposure was not associated with endotoxin, even though dog ownership was a strong predictor of this GNB biomarker in a previous analysis of this cohort [8, 10]. It is possible that more detailed questions about the dog's activity (where the dog was permitted inside the household, whether it often went outside) would have refined this home characteristic, lending more power to detect a link between dogs and house dust endotoxin in the present work. Cat allergen levels were associated with bed dust endotoxin, a finding consistent with other studies [26]. However, cat ownership was not associated with increased GNB, suggesting that cat allergen measurement is actually a better indicator of pet presence in the sampling location. It may be better because a) allergen may be brought into the household even when the participants do not own the pet, or b) allergen and microbes associated with pets may vary depending on whether or not the pets are allowed in the bedroom. Nevertheless, report of a pet is generally closely correlated with high pet allergen levels, whereas (as we have previously reported) sightings of pests are poorly correlated with allergen levels for either cockroach or mice [27].

Next to sources of moisture, cleaning frequency was the most consistent variable associated with the microbial environment in the home. Cleaning the bedroom at least once a week was related to decreased levels of fungi, GPB, and GNB. Lower cleaning frequency (vacuuming and dusting) has been associated with increased levels of β-d-glucan and 3-OHFAs in other reports [22, 23, 16]. The potential for an individual to alter his or her home microbial exposure, perhaps as a result of experiencing allergy symptoms, should be accounted for in epidemiologic studies on home exposures and allergic disease. Because an individual s own commensal flora has been shown to affect the composition of house dust [28] and could potentially vary by sex, we examined sex as a predictor in multiple regression models. Although sex was not associated with microbial biomarker levels in the bedroom, the number of boys was positively associated with bacteria and fungal markers in the family room, perhaps because of differential behaviors by sex [29].

In this cohort, low-income level was linked to decreased levels of fungal exposure (ergosterol). Race/ethnicity was associated with reductions in GNB (endotoxin) and fungi (β-d-glucan), with black index children showing the lowest home exposures to these microbes. The race/ethnicity variable remained a significant predictor of reduced microbial levels even after adjustment for living in an apartment and income level. These associations are most likely driven by differences in housing and living conditions by race/ethnicity that are not captured in the home characteristics questionnaire. It is likely that the relationship of ethnicity to levels of home microbial levels can be attributed to residual confounding by unmeasured housing characteristics (e.g., control over home heating) that differ by race/ethnicity. An inverse association between black race/ethnicity and endotoxin level was also observed in an earlier analysis of this birth cohort, relating endotoxin exposure in the first year of life with home/demographic characteristics [10, 17]. The relationship between microbial exposure levels and SES may, at least in part, account for the higher rates of respiratory disease observed in minority populations. Because endotoxin protects against allergic disease in our cohort, the lower GNB levels in African-American households may represent unmeasured home characteristics that represent a risk factor for allergy and asthma.

Our epidemiologic approach could also have translational implications. Our main focus was not on pathogenic microorganisms that may cause clusters of acute disease. In our research we use microbial markers for epidemiologic reasons, with more of an interest in exposure to elevated levels of commensal organisms. Many commensal microbial organisms have irritant properties. However, exposure to microbial components at critical stages of life is also hypothesized to protect against allergic disease. The more specifically we identify organisms adversely or positively associated with health effects, the more we can do translational research, with the potential to identify components that might ultimately have therapeutic purposes. This is an additional motivating factor for measuring specific biomarkers in the home, rather than relying on questionnaire responses.

Predictive Value of Home Characteristics

We focused on the predictive power of home characteristics for levels of biomarkers above the median because we have identified health effects of exposures categorized using this cut-point [10, 30]. PPVs of home characteristics were generally poor, although dampness, infrequent cleaning, and pets predicted elevated (> median) levels of GNB in bed dust samples 71–75% of the time. However, we saw a drop in these PPVs and an increase in the NPVs when we set the threshold for elevated microbial levels at the 75th percentile (as opposed to the median). Thus, the absence of dampness, high cat allergen, or infrequent cleaning predicts the absence of higher mold or GPB at this cut-point.

To use home characteristics as exposure surrogates with minimal misclassification, both PPV and NPV values would have to be sufficiently high (≥ 90%). Although shifting the microbial exposure threshold produced increases in either PPV or NPV, these values never simultaneously reached the predictive probabilities necessary to justify the use of home characteristics as substitutes for microbial biomarker quantification.

Limitations

Although we were able to study predictors of the complex microbial milieu at a single time point, measurement of the entire biomarker panel at additional time points was cost prohibitive, thereby limiting our ability to analyze longitudinal changes in home microbial levels. We assessed cleaning frequency in the bedroom by questionnaire, but we did not have data on specific cleaning practices (e.g., dry vs. wet mopping/dusting) that might be associated with decreasing microbial biomarker levels. Another limitation of this study is the sensitivity of predictive value calculations to changes in cut-points for "high" microbial exposure levels. For this reason, our main statistical analyses focused on microbial biomarkers as continuous outcome measures, which yielded the most stable estimate.

CONCLUSIONS

Significant predictors of home microbial exposure encompass categories of dampness, pets, cleaning, and demographics (SES, race/ethnicity, and sex). Home characteristics partially explain variation in microbial exposure levels, but they cannot serve as surrogates or as markers of differentiation between microbes. In this U.S. urban environment, if one is seeking to evaluate health effects of specific groups of organisms that (like mold vs. GNB) may have differing effects on children [31], then it becomes necessary to measure those biomarkers rather than relying solely on home characteristics to profile microbial exposures in the home.

KEYWORDS

- **Bacteria**
- **Dampness**
- **Fungi**
- **Home characteristics**
- **Indoor exposure**
- **Respiratory health**

REFERENCES

1. Douwes J, van der Sluis B, Doekes G, van Leusden F, Wijnands L, van Strien R, et al. Fungal extracellular polysaccharides in house dust as a marker for exposure to fungi: relations with culturable fungi, reported home dampness, and respiratory symptoms. J Allergy Clin Immunol. 1999; 103(3 pt 1):494–500.
2. Fung F, Hughson WG. Health effects of indoor fungal bioaerosol exposure. Applied ccupational Environmental Hygeine 2003; 18(7):535–544.
3. Garrett MH, Rayment PR, Hooper MA, Abramson MJ, Hooper BM. Indoor airborne fungal spores, house dampness and associations with environmental factors and respiratory health in children. Clin Exp Allergy 1998; 28(4):459–467.
4. Handal G, Leiner MA, Cabrera M, Straus DC. Children symptoms before and after knowing about an indoor fungal contamination. Indoor Air 2004; 14(2):87–91.
5. Stark PC, Celedon JC, Chew GL, Ryan LM, Burge HA, Mulienberg ML, et al. Fungal levels in the home and allergic rhinitis by 5 years of age. Environmental Health Perspectives 2005; 113:1405–1409.
6. Celedon JC, Milton DK, Ramsey CD, Litonjua AA, Ryan L, Platts-Mills TA, et al. Exposure to dust mite allergen and endotoxin in early life and asthma and atopy in childhood. Journal of Allergy Clinical Immunology 2007; 120:144–149.
7. Litonjua AA, Milton DK, Celedon JC, Ryan L, Weiss ST, Gold DR. A longitudinal analysis of wheezing in young children: the independent effects of early life exposure to house dust endotoxin, allergens, and pets. Journal of Allergy Clinical Immunology 2002; 110(5):736–742.
8. Park JH, Spiegelman DL, Gold DR, Burge HA, Milton DK. Predictors of airborne endotoxin in the home. Environmental Health Perspectives 2001b; 109:859–864.
9. Gold DR, Burge HA, Carey V, Milton DK, Platts-Mills T, Weiss ST. Predictors of repeated wheeze in the first year of life: the relative roles of cockroach, birth weight, acute lower respiratory illness, and maternal smoking. American Journal Respiratory Critical Care Medicine 1999; 160(1):227–236.
10. Park JH, Gold DR, Spiegelman DL, Burge HA, Milton DK. House dust endotoxin and wheeze in the first year of life. American Journal of Respiratory Critical Care Medicine 2001a; 163(2):322–328.
11. Park JH, Szponar B, Larsson L, Gold DR, Milton DK. Characterization of lipopolysaccharides present in settled house dust. Applied Environmental Microbiology 2004; 70(1):262–267.
12. Sebastian A, Harley W, Fox A, Larsson L. Evaluation of the methyl ester O-methyl acetate derivative of muramic acid for the determination of peptidoglycan in environmental samples by ion-trap GC-MS-MS. Journal of Environmental Monitor 2004; 6(4):300–304.
13. Milton DK, Alwis KU, Fisette L, Muilenberg M. Enzyme-linked immunosorbent assay specific for (1→6) branched, (1→3)-beta-d-glucan detection in environmental samples. Applied Environmental Microbiology 2001; 67(12):5420–5424.
14. Axelsson BO, Saraf A, Larsson L. Determination of ergosterol in organic dust by gas chromatography mass spectrometry. J Chromatogr B Biomed Appl. 1995; 666(1):77 84.
15. Sebastian A, Larsson L. Characterization of the microbial community in indoor environments: a chemical-analytical approach. Applied Environmental Microbiology 2003; 69(6):3103–3109.
16. Hyvarinen A, Sebastian A, Pekkanen J, Larsson L, Korppi M, Putus T, et al. Characterizing microbial exposure with ergosterol, 3-hydroxy fatty acids, and viable microbes in house dust: determinants and association with childhood asthma. Archives of Environvental Occupational Health 2006; 61(4):149–157.
17. Park JH, Spiegelman DL, Burge HA, Gold DR, Chew GL, Milton DK. Longitudinal study of dust and airborne endotoxin in the home. Environmental Health Perspectives 2000; 108:1023–1028.
18. Wickens K, Douwes J, Siebers R, Fitzharris P, Wouters I, Doekes G, et al. Determinants of endotoxin levels in carpets in New Zealand homes. Indoor Air 2003; 13(2):128–135.

19. Gereda JE, Klinnert MD, Price MR, Leung DY, Liu AH. Metropolitan home living conditions associated with indoor endotoxin levels. Journal of Allergy Clinical Immunology 2001; 107(5):790–796.
20. O'Connor GT, Walter M, Mitchell H, Kattan M, Morgan WJ, Gruchalla RS, et al. Airborne fungi in the homes of children with asthma in low-income urban communities: the Inner-City Asthma Study. Journal of Allergy Clinical Immunology 2004; 114(3):599–606.
21. Nevalainen A, Seuri M. Of microbes and men. Indoor Air. 2005; 15(suppl 9):58–64.
22. Bischof W, Koch A, Gehring U, Fahlbusch B, Wichmann HE, Heinrich J. Predictors of high endotoxin concentrations in the settled dust of German homes. Indoor Air 2002; 12(1):2–9.
23. Gehring U, Douwes J, Doekes G, Koch A, Bischof W, Fahlbusch B, et al. Beta(1 → 3)-glucan in house dust of German homes: housing characteristics, occupant behavior, and relations with endotoxins, allergens, and molds. Environmental Health Perspectives 2001; 109:139–144.
24. Heinrich J, Gehring U, Douwes J, Koch A, Fahlbusch B, Bischof W, et al. Pets and vermin are associated with high endotoxin levels in house dust. Clin Exp Allergy 2001; 31(12):1839–1845.
25. Thorne PS, Cohn RD, Mav D, Arbes SJ, Zeldin DC. Predictors of endotoxin levels in U.S. housing. Environmental Health Perspectives 2009; 117:763–771.
26. Giovannangelo M, Gehring U, Nordling E, Oldenwening M, Terpstra G, Bellander T, et al. Determinants of house dust endotoxin in three European countries—the AIRALLERG study. Indoor Air 2007; 17(1):70–79.
27. Chew GL, Burge HA, Dockery DW, Muilenberg ML, Weiss ST, Gold DR. Limitations of a home characteristics questionnaire as a predictor of indoor allergen levels. American Journal of Respir Crit Care Medicine 1998; 157:1536–1541.
28. Täubel M, Rintala H, Pitkäranta M, Paulin L, Laitinen S, Pekkanen J, et al. The occupant as a source of house dust bacteria. Journal of Allergy Clinical Immunology 2009; 124(4):834–840.
29. Else-Quest NM, Hyde JS, Goldsmith HH, Van Hulle CA. Gender differences in temperament: a meta-analysis. Psychol Bull. 2006; 132(1):33–72.
30. Litonjua AA, Milton DK, Celedon JC, Ryan L, Weiss ST, Gold DR. A longitudinal analysis of wheezing in young children: the independent effects of early life exposure to house dust endotoxin, allergens, and pets. Journal of Allergy Clinical Immunology 2002; 110(5):736–742.
31. Zeldin DC, Eggleston P, Chapman M, Piedimonte G, Renz H, Peden D. How exposures to biologics influence the induction and incidence of asthma. Environmental Health Perspectives 2006; 114:620–626.

5 Pollutants from Vehicle Exhaust Near Highways

Doug Brugge, John L. Durant, and Christine Rioux

CONTENTS

INTRODUCTION

There is growing evidence of a distinct set of freshly emitted air pollutants downwind from major highways, motorways, and freeways that include elevated levels of ultrafine particulates (UFP), black carbon (BC), oxides of nitrogen (NOx), and carbon monoxide (CO). People living or otherwise spending substantial time within about 200 m of highways are exposed to these pollutants more so than persons living at a greater distance, even compared to living on busy urban streets. Evidence of the health hazards of these pollutants arises from studies that assess proximity to highways, actual exposure to the pollutants, or both. Taken as a whole, the health studies show elevated risk for development of asthma and reduced lung function in children who live near major highways. Studies of particulate matter (PM) that show associations with cardiac and pulmonary mortality also appear to indicate increasing risk as smaller geographic areas are studied, suggesting localized sources that likely include major highways. Although less work has tested the association between lung cancer and highways, the existing studies suggest an association as well. While the evidence is substantial for a link between near-highway exposures and adverse health outcomes, considerable work remains to understand the exact nature and magnitude of the risks.

BACKGROUND

Approximately 11% of US households are located within 100 meters of 4-lane highways [estimated using: [1, 2]]. While it is clear that automobiles are significant sources of air pollution, the exposure of near-highway residents to pollutants in automobile exhaust has only recently begun to be characterized. There are two main reasons for this: (A) federal and state air monitoring programs are typically set up to measure pollutants at the regional, not local scale; and (B) regional monitoring stations typically do not measure all of the types of pollutants that are elevated next to highways. It is, therefore, critical to ask what is known about near-highway exposures and their possible health consequences.

Here we review studies describing measurement of near-highway air pollutants, and epidemiologic studies of cardiac and pulmonary outcomes as they relate to exposure to these pollutants and/or proximity to highways. Although some studies suggest that other health impacts are also important (e.g., birth outcomes), we feel that the case for these health effects are less well developed scientifically and do not have the same potential to drive public policy at this time. We did not seek to fully integrate the relevant cellular biology and toxicological literature, except for a few key references, because they are so vast by themselves.

We started with studies that we knew well and also searched the engineering and health literature on Medline. We were able to find some earlier epidemiologic studies based on citations in more recent articles. We include some studies that assessed motor vehicle-related pollutants at central site monitors (i.e., that did not measure highway proximity or traffic) because we feel that they add to the plausibility of the associations seen in other studies. The relative emphasis given to studies was based on our appraisal of the rigor of their methodology and the significance of their findings. We conclude with a summary and with recommendations for policy and further research.

MOTOR VEHICLE POLLUTION

It is well known that motor vehicle exhaust is a significant source of air pollution. The most widely reported pollutants in vehicular exhaust include carbon monoxide, nitrogen and sulfur oxides, unburned hydrocarbons (from fuel and crankcase oil), particulate matter, polycyclic aromatic hydrocarbons, and other organic compounds that derive from combustion [3–5]. While much attention has focused on the transport and transformation of these pollutants in ambient air—particularly in areas where both ambient pollutant concentrations and human exposures are elevated (e.g., congested city centers, tunnels, and urban canyons created by tall buildings), less attention has been given to measuring pollutants and exposures near heavily-trafficked highways. Several lines of evidence now suggest that steep gradients of certain pollutants exist next to heavily traveled highways and that living within these elevated pollution zones can have detrimental effects on human health.

It should be noted that many different types of highways have been studied, ranging from California "freeways" (defined as multi-lane, high-speed roadways with restricted access) to four-lane (two in each direction), variable-speed roadways with unrestricted access. There is considerable variation in the literature in defining highways and we choose to include studies in our review that used a broad range of definitions (see Table 1).

It should also be noted that there may be significant heterogeneity in the types and amounts of vehicles using highways. The typical vehicle fleet in the US is composed of passenger cars, sports utility vehicles, motorcycles, pickup trucks, vans, buses, and small, medium, and large trucks. The composition and size of a fleet on a given highway may vary depending on the time of day, day of the week, and use restrictions for certain classes of vehicles. Fleets may also vary in the average age and state of repair of vehicles, the fractions of vehicles that burn diesel and gasoline, and the fraction of vehicles that have catalytic converters. These factors will influence the kinds and amounts of pollutants in tailpipe emissions. Similarly, driving conditions, fuel chemistry, and meteorology can also significantly impact emissions rates as well as the kinds and concentrations of pollutants present in the near-highway environment. These factors have rarely been taken into consideration in health outcome studies of near-highway exposure.

Based on our review of the literature, the pollutants that have most consistently been reported at elevated levels near highways include ultrafine particles (UFP), black carbon (BC), nitrogen oxides (NOx), and carbon monoxide (CO). In addition, $PM_{2.5}$, and PM_{10} were measured in many of the epidemiologic studies we reviewed. UFP are defined as particles having an aerodynamic diameter in the range of 0.005 to 0.1 microns (um). UFP form by condensation of hot vapors in tailpipe emissions, and can grow in size by coagulation. $PM_{2.5}$ and PM_{10} refer to particulate matter with aerodynamic diameters of 2.5 and 10 um, respectively. BC (or "soot carbon") is an impure form of elemental carbon that has a graphite-like structure. It is the major light-absorbing component of combustion aerosols. These various constituents can be measured in real time or near-real time using particle counters (UFP) and analyzers that measure light absorption (BC and CO), chemiluminescence (NOx), and weight ($PM_{2.5}$ and PM_{10}). Because UFP, NO_x, BC, and CO derive from a common source—vehicular emissions—they are typically highly inter-correlated.

AIR POLLUTANT GRADIENTS NEAR HIGHWAYS

Several recent studies have shown that sharp pollutant gradients exist near highways. Shi et al. [6] measured UFP number concentration and size distribution along a roadway-to-urban-background transect in Birmingham (UK), and found that particle number concentrations decreased nearly 5-fold within 30 m of a major roadway (>30, 000 veh/d). Similar observations were made by Zhu et al. [7, 8] in Los Angeles. Zhu et al. measured wind speed and direction, traffic volume, UFP number concentration and size distribution as well as BC and CO along transects downwind of a highway that is dominated by gasoline vehicles (Freeway 405; 13, 900 vehicles per hour; veh/h) and a highway that carries a high percentage of diesel vehicles (Freeway 710; 12, 180 veh/h). Relative concentrations of CO, BC, and total particle number concentration

TABLE 1 Summary of Near-Highway Pollution Gradients.

Citation	Location	Highway traffic intensity[a]	Pollutants measured[b]	Observed Pollution Gradients
Shi et al. 1999 (6)	Birmingham, UK	30, 000 veh/d	UFP + FP (10-10^4 nm)	2–100 m [c]
Zhu et al. 2002 (8)	Los Angeles; Freeway 710	12, 180 veh/h	UFP, CO, BC	17–300 m [c]
Zhu et al. 2002 (7)	Los Angeles; Freeway 405	13, 900 veh/h	UFP, CO, BC	30–300 m [c]
Hitchins et al. 2002 (11)	Brisbane (Austr.)	2, 130–3, 400 veh/h	UFP + FP (15-2 × 10^4 nm), $PM_{2.5}$	15–375 m [c]
Fischer et al. 2000 (13)	Amsterdam	<3, 000–30, 974 veh/d	$PM_{2.5}$, PM_{10}, PPAH, VOCs	NA
Roorda-Knape et al. 1998 (14)	Netherlands	80, 000–152, 000 veh/d	$PM_{2.5}$, PM_{10}, BC, VOCs, NO_2	15–330 m [c]
Janssen et al. 2001 (15)	Netherlands	40, 000–170, 000 veh/d	$PM_{2.5}$, VOCs, NO_2	< 400 m [c]
Morawska et al. 1999 (12)	Brisbane (Austr.)	NA	UFP	10–210 m [c]

[a]As defined in article cited (veh/d = vehicles per day; veh/h = vehicles per hour).
[b]UFP = ultrafine particles; FP = fine particles; $PM_{2.5}$ = particles with aerodynamic diameter ≤ 2.5 um; PM_{10} = particles with aerodynamic diameter ≤ 10 um; BC = black carbon; PPAH = particle-bound polycyclic aromatic hydrocarbons; VOCs = volatile organic compounds
[c]Pollutant measurements were made along a transect away from the highway
NA = not applicable; measurements were not made.

decreased exponentially between 17 and 150 m downwind from the highways, while at 300 m UFP number concentrations were the same as at upwind sites. An increase in the relative concentrations of larger particles and concomitant decrease in smaller particles was also observed along the transects (see Figure 1).

Similar observations were made by Zhang et al. [9] who demonstrated "road-to-ambient" evolution of particle number distributions near highways 405 and 710 in both winter and summer. Zhang et al. observed that between 30–90 m downwind of the highways, particles grew larger than 0.01 um due to condensation, while at distances >90 m, there was both continued particle growth (to >0.1 um) as well as particle shrinkage to <0.01 um due to evaporation. Because condensation, evaporation, and dilution alter size distribution and particle composition, freshly emitted UFP near highways may differ in chemical composition from UFP that has undergone atmospheric transformation during transport to downwind locations [10].

Two studies in Brisbane (Australia) highlight the importance of wind speed and direction as well as contributions of pollutants from nearby roadways in tracking highway-generated pollutant gradients. Hitchins et al. [11] measured the mass concentrations of 0.1–10 um particles as well as total particle number concentration and size distribution for 0.015–0.7 um particles near highways (2, 130–3, 400 veh/h). Hitchens et al. observed that the distance from highways at which number and mass concentrations decreased by 50% varied from 100 to 375 m depending on the wind speed and direction. Morawska et al. [12] measured the changes in UFP number concentrations along horizontal and vertical transects near highways to distinguish highway and normal street traffic contributions. It was observed that UFP number concentrations were highest <15 m from highways, while 15–200 m from highways there was no significant difference in UFP number concentrations along either horizontal or vertical transects—presumably due to mixing of highway pollutants with emissions from traffic on nearby, local roadways.

In addition to UFP, other pollutants—such as $PM_{2.5}$, PM_{10}, NO_2 (nitrogen dioxide), VOCs (volatile organic compounds), and particle-bound polycyclic aromatic hydrocarbons (PPAH)—have been studied in relation to heavily trafficked roadways. Fischer et al. [13] measured $PM_{2.5}$, PM_{10}, PPAH, and VOC concentrations outside and inside homes on streets with high and low traffic volumes in Amsterdam (<3, 000–30, 974 veh/d). In this study, PPAH and VOCs were measured using methods based on gas chromatography. Fischer et al. found that while $PM_{2.5}$ and PM_{10} mass concentrations were not specific indicators of traffic-related air pollution, PPAH and VOC levels were ~2-fold higher both indoor and outdoor in high traffic areas compared to low traffic areas. Roorda-Knape et al. [14] measured $PM_{2.5}$, PM_{10}, black smoke (which is similar to BC), NO_2, and benzene in residential areas <300 m from highways (80,000–152, 000 veh/d) in the Netherlands. Black smoke was measured by a reflectance-based method using filtered particles; benzene was measured using a method based on gas chromatography. Roorda-Knape et al. reported that outdoor concentrations of black smoke and NO_2 decreased with distance from highways, while $PM_{2.5}$, PM_{10}, and benzene concentrations did not change with distance. In addition, Roorda-Knape et al. found that indoor black smoke concentrations were correlated with truck traffic, and NO_2 was correlated with both traffic volume and distance from highways. Janssen et al.

[15] studied $PM_{2.5}$, PM_{10}, benzene, and black smoke in 24 schools in the Netherlands and found that $PM_{2.5}$ and black smoke increased with truck traffic and decreased with distance from highways (40, 000–170, 000 veh/d).

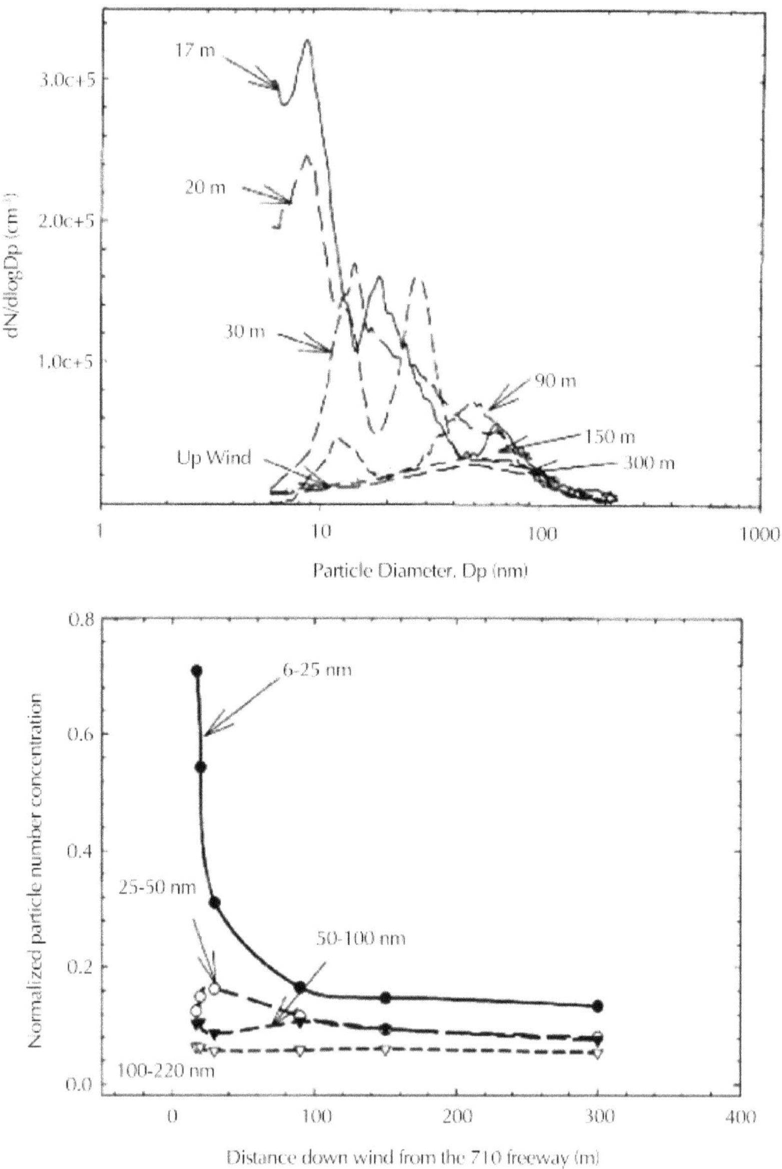

FIGURE 1 Ultrafine particle size distribution (top panel) and normalized particle number concentration for different size ranges (bottom panel) as a function of distance from a highway in Los Angeles. From Zhu et al. [8]. Reprinted with permission from Elsevier.

In summary, the literature shows that UFP, BC, CO and NOx are elevated near highways (>30, 000 veh/d), and that other pollutants including VOCs and PPAHs may also be elevated. Thus, people living within about 30 m of highways are likely to receive much higher exposure to traffic-related air pollutants compared to residents living >200 m (+/- 50 m) from highways.

CARDIOVASCULAR HEALTH AND TRAFFIC-RELATED POLLUTION

Results from clinical, epidemiological, and animal studies are converging to indicate that short-term and long-term exposures to traffic-related pollution, especially particulates, have adverse cardiovascular effects [16–18]. Most of these studies have focused on, and/or demonstrated the strongest associations between cardiovascular health outcomes and particulates by weight or number concentrations [19–21] though CO, SO_2, NO_2, and BC have also been examined. BC has been shown to be associated with decreases in heart rate variability (HRV) [22, 23] and black smoke and NO_2 shown to be associated with cardiopulmonary mortality [24].

Short-term exposure to fine particulate pollution exacerbates existing pulmonary and cardiovascular disease and long-term repeated exposures increases the risk of cardiovascular disease and death [25, 26].

Though not focused on near-highway pollution, two large prospective cohort studies, the Six-Cities Study [27] and the American Cancer Society (ACS) Study [28] provided the groundwork for later research on fine particulates and cardiovascular disease. Both of these studies found associations between increased levels of exposure to ambient PM and sulfate air pollution recorded at central city monitors and annual average mortality from cardiopulmonary disease, which at the time combined cardiovascular and pulmonary disease other than lung cancer. The Six-Cities Study examined $PM_{2.5}$ and $PM_{10/15}$. The ACS study examined $PM_{2.5}$. Relative risk ratios of mortality from cardiopulmonary disease comparing locations with the highest and lowest fine particle concentrations (which had differences of 24.5 and 18.6 ug/m^3 respectively) were 1.37 (1.11, 1.68) and 1.31 (1.17, 1.46) in the Six Cities and ACS studies, respectively. These analyses controlled for many confounders, including smoking and gas stoves but not other housing conditions or time spent at home. The studies were subject to intensive replication, validation, and reanalysis that confirmed the original findings. $PM_{2.5}$ generally declined following implementation of new U.S. Environmental Protection Agency standards in 1997 [17, 29], yet since that time studies have shown elevated health risks due to long-term exposures to the 1997 PM threshold concentrations [29, 30].

Much of the epidemiological research has focused on assessing the early physiological responses to short-term fluctuations in air pollution in order to understand how these exposures may alter cardiovascular risk profiles and exacerbate cardiovascular disease [31]. Heart rate variability, a risk factor for future cardiovascular outcomes, is altered by traffic-related pollutants particularly in older people and people with heart disease [22, 23, 32]. With decreased heart rate variability as the adverse outcome, negative associations between HRV and particulates were strongest for the smallest size fraction studied [33] (PM0.3–1.0); [34] (PM0.02–1). In two studies that included other pollutants, black carbon, an indicator of traffic particles, also elicited a strong

TABLE 2 Summary of Near-Highway Health Effects Studies.

Citation	Location	Highway traffic intensity[a]	Pollutants measured[b]	Distance from highway	Health Outcomes	Statistical association[c]
Schwartz et al. 2005 (22)	Boston	NA	$PM_{2.5}$, BC, CO	NA	Heart rate variability	Decreases in measures of heart rate variability
Adar et al. 2007 (23)	St. Louis, Missouri	NA	$PM_{2.5}$, BC, UFP	On highway in busses	Heart rate variability	Decreases in measures of heart rate variability
Hoek et al. 2002 (24)	Netherlands	NA	BC, NO_2	Continuous [d]	Cardio-pulmonary mortality, lung cancer	1.41 OR for living near road
Tonne et al. 2007 (41)	Worchester, Mass.	NA	$PM_{2.5}$	Continuous [d]	Acute myocardial infarction (AMI)	5% increase in odds of AMI
Venn et al. 2001 (49)	Nottingham, UK	NA	NA	Continuous [d]	Wheezing in children	1.08 OR for living w/in 150 m of road
Nicolai et al. 2003 (58)	Munich, Germany	>30, 000 veh/d	Soot, benzene, NO_2	Traffic counts within 50 m of house	Asthma, respiratory symptoms, allergy	1.79 OR for asthma and high traffic volume
Gauderman et al. 2005 (65)	Southern California		NO_2	Continuous [d]	Asthma, respiratory symptoms	Increased asthma closer to freeways
McConnell et al. 2006 (57)	Southern California	NA	NA	Continuous [d]	Asthma	Large risk for children living w/in 75 m of road
Ryan, et al. 2007 (59)	Cincinnati, Ohio	> 1, 000 trucks/d	PM2.5	400 m	Wheezing in children	NA
Kim et al. 2004 (60)	San Francisco	90, 000 – 210, 000 veh/d	PM, BC, NO_x	School sites	Childhood asthma	1.07 OR for high levels of NO_x

TABLE 2 *(Continued)*

Citation	Location	Highway traffic intensity[a]	Pollutants measured[b]	Distance from highway	Health Outcomes	Statistical association[e]
Wjst et al. 1993 (68)	Munich, Germany	7, 000–125, 000 veh/d	NO_x, CO	School sites	Asthma, bronchitis	Several statistical associations found
Brunekreef et al. 1997 (69)	Netherlands	80, 000 – 152, 000 veh/d	PM_{10}, NO_2	Continuous[d]	Lung function	Decreased FEV with proximity to high truck traffic
Janssen et al. 2003 (74)	Netherlands	30, 000–155, 000 veh/d	$PM_{2.5}$, NO_2, benzene	< 400 m [c]	Lung function, respiratory symptoms	No association with lung function
Peters et al. 1999 (82)	Southern California	NA	PM_{10}, NO_2	NA	Asthma, bronchitis, cough, wheeze	1.54 OR of wheeze for boys with exposure to NO_2
Brauer et al. 2007 (67)	Netherlands	Highways and streets	$PM_{2.5}$, NO_2, soot	Modeled exposure	Asthma, allergy, bronchitis, respiratory symptoms	Strongest association was with food allergies
Visser et al. 2004 (91)	Amsterdam	> 10, 000 veh/d	NA	NA	Cancer	Multiple associations
Vineis et al. 2006 (87)	10 Eurpoean countries	NA	PM_{10}, NO_2, SO_2	NA	Cancer	1.46 OR near heavy traffic, 1.30 OR for high exposure to NO_2
Gauderman et al. 2007 (73)	Southern California	NA	PM_{10}, NO_2	Continuous[d]	Lung Function	Decreased FEV for those living near freeway

[a]As defined in article cited (veh/d = vehicles per day; veh/h = vehicles per hour).
[b]UFP = ultrafine particles; FP = fine particles; $PM_{2.5}$ = particles with aerodynamic diameter ≤ 2.5 um; PM_{10} = particles with aerodynamic diameter ≤ 10 um; BC = black carbon; PPAH = particle-bound polycyclic aromatic hydrocarbons; VOCs = volatile organic compounds
[c]Pollutant measurements were made along a transect away from the highway
[d]Proximity of each participant to a major road was calculated using GIS software
[e]Statistical association between proximity to highway or exposure to traffic-generated pollutants and measured health outcomes
NA = not applicable; measurements were not made.

association with both time and frequency domain HRV variables; associations were also strong for $PM_{2.5}$ for both time and frequency HRV variables in the Adar et al study [23]; this and subsequent near highway studies are summarized in Table 2], however, $PM_{2.5}$ was not associated with frequency domain variables in the Schwartz et al. study [22].

Several studies show that exposure to PM varies spatially within a city [35–37], and finer spatial analyses show higher risks to individuals living in close proximity to heavily trafficked roads [18, 37]. A 2007 paper from the Women's Health Initiative used data from 573 $PM_{2.5}$ monitors to follow over 65, 000 women prospectively. They reported very high hazard ratios for cardiovascular events (1.76; 95% CI, 1.25 to 2.47) possibly due to the fine grain of exposure monitoring [18]. In contrast, studies that relied on central monitors [27, 28] or interpolations from central monitors to highways are prone to exposure misclassification because individuals living close to highways will have a higher exposure than the general area. A possible concern with this interpretation is that social gradients may also situate poorer neighborhoods with potentially more susceptible populations closer to highways [38–40].

At a finer grain, Hoek et al. [24] estimated home exposure to nitrogen dioxide (NO_2) and black smoke for about 5, 000 participants in the Netherlands Cohort Study on Diet and Cancer. Modeled exposure took into consideration proximity to freeways and main roads (100 m and 50 m, respectively). Cardiopulmonary mortality was associated with both modeled levels of pollutants and living near a major road with associations less strong for background levels of both pollutants. A case-control study [41], found a 5% increase in acute myocardial infarction associated with living within 100 m of major roadways. A recent analysis of cohort data found that traffic density was a predictor of mortality more so than was ambient air pollution [42]. There is a need for studies that assess exposure at these scales, e.g., immediate vicinity of highways, to test whether cardiac risk increases still more at even smaller scales.

Although we cannot review it in full here, we note that evidence beyond the epidemiological literature support the contention that $PM_{2.5}$ and UFP (a sub-fraction of $PM_{2.5}$) have adverse cardiovascular effects [16, 17]. $PM_{2.5}$ appears to be a risk factor for cardiovascular disease via mechanisms that likely include pulmonary and systemic inflammation, accelerated atherosclerosis and altered cardiac autonomic function [17, 22, 43–46]. Uptake of particles or particle constituents in the blood can affect the autonomic control of the heart and circulatory system. Black smoke, a large proportion of which is derived from mobile source emissions [30], has a high pulmonary deposition efficiency, and due to their surface area-to-volume ratios can carry relatively more adsorbed and condensed toxic air pollutants (e.g., PPAH) compared to larger particles [17, 47, 48]. Based on high particle numbers, high lung deposition efficiency and surface chemistry, UFP may provide a greater potential than $PM_{2.5}$ for inducing inflammation [10]. UFPs have high cytotoxic reactive oxygen species (ROS) activity, through which numerous inflammatory responses are induced, compared to other particles [10]. Chronically elevated UFP levels such as those to which residents living near heavily trafficked roadways are likely exposed can lead to long-term or repeated increases in systemic inflammation that promote arteriosclerosis [18, 29, 34, 37].

ASTHMA AND HIGHWAY EXPOSURES

Evidence that near highway exposures present elevated risk is relatively well developed with respect to child asthma studies. These studies have evolved over time with the use of different methodologies. Studies that used larger geographic frames and/or overall traffic in the vicinity of the home or school [49–52] or that used self-report of traffic intensity [53] found no association with asthma prevalence. Most recent child asthma studies have, instead, used increasingly narrow definitions of proximity to traffic, including air monitoring or modeling) and have focused on major highways instead of street traffic [54–59]. All of these studies have found statistically significant associations between the prevalence of asthma or wheezing and living very close to high volume vehicle roadways. Confounders considered included housing conditions (pests, pets, gas stoves, water damage), exposure to tobacco smoke, various measures of socioeconomic status (SES), age, sex, and atopy, albeit self-reported and not all in a single study.

Multiple studies have found girls to be at greater risk than boys for asthma resulting from highway exposure [55, 57, 60]. A recent study also reports elevated risk only for children who moved next to the highway before they were 2 years of age, suggesting that early childhood exposure may be key [57]. The combined evidence suggests that living within 100 meters of major highways is a risk factor, although smaller distances may also result in graded increases in risk. The neglect of wind direction and the absence of air monitoring from some studies are notable missing factors. Additionally, recent concerns have been raised that geocoding (attaching a physical location to addresses) could introduce bias due to inaccuracy in locations [61].

Studies that rely on general area monitoring of ambient pollution and assess regional pollution on a scale orders of magnitude greater than the near-roadway gradients have also found associations between traffic generated pollution (CO and NOx) and prevalence of asthma [62] or hospital admission for asthma [63]. Lweguga-Mukasa et al. [64] monitored air up and down wind of a major motor vehicle bridge complex in Buffalo, NY and found that UFP were higher downwind, dropping off with distance. Their statistical models did not, however, support an association of UFP with asthma. A study in the San Francisco Bay Area measured $PM_{2.5}$, BC and NOX over several months next to schools and found both higher pollution levels downwind from highways and a linear association of BC with asthma in long-term residents [60].

Gauderman et al. [65] measured NO_2 next to homes of 208 children. They found an odds ratio (OR) of 1.83 (confidence interval (CI): 1.04–3.22) for outdoor NO_2 (probably a surrogate for total highway pollution) and lifetime diagnosis of asthma. They also found a similar association with distance from residence to freeway. Self-report was used to control for numerous confounders, including tobacco smoke, SES, gas stoves, mildew, water damage, cockroaches and pets, which did not substantially affect the association. Gauderman's study suggests that ambient air monitoring at the residence substantially increases statistical power to detect association of asthma with highway exposures.

Modeling of elemental carbon attributable to traffic near roadways based on ambient air monitoring of $PM_{2.5}$ has recently emerged as a viable approach and a study using this method found an association with infant wheezing. The modeled values appear to be better predictors than proximity. Elevation of the residence relative t traffic was also an important factor in this study [66]. A 2007 paper reported on mod eled NO_2, $PM_{2.5}$ and soot and the association of these values with asthma and vari ous respiratory symptoms in the Netherlands [67]. While finding modest statisticall significant associations for asthma and symptoms, it is somewhat surprising that the found stronger associations for development of sensitization to food allergens.

PEDIATRIC LUNG FUNCTION AND TRAFFIC-RELATED AIR POLLUTION

Studies of association of children's lung function with traffic pollutants have used a variety of measures of exposure, including: traffic density, distance to roadways, are (city) monitors, monitoring at the home or school and personal monitoring. Studie have assessed both chronic effects on lung development and acute effects and hav been both cross-sectional and longitudinal. The wide range of approaches somewha complicates evaluation of the literature.

Traffic density in school districts in Munich was associated with decreases i forced vital capacity (FVC), forced expiratory volume in 1 second (FEV1), FEV1 FVC and other measures, although the 2-kilometer (km) areas, the use of sitting po sition for spirometry and problems with translation for non-German children wer limitations [68]. Brunekreef et al. [69] used distance from major roadways, considere wind direction and measured black smoke and NO_2 inside schools. They found th largest decrements in lung function in girls living within 300 m of the roadways.

A longitudinal study of children (average age at start = 10 years) in Souther California reported results at 4 [70] and 8 years [71]. Multiple air pollutants were mea sured at sites in 12 communities. Due to substantial attrition, only 42% of children en rolled at the start were available for the 8-year follow-up. Substantially lower growt in FEV1 was associated with PM_{10}, NO_2, $PM_{2.5}$, acid vapor and elemental carbon at 4 and at 8 years. The analysis could not indicate whether the effects seen were reversibl or not [72]. In 2007, it was reported from this same cohort that living within 500 m o a freeway was reported to be associated with reduced lung function [73].

A Dutch study [74] measured $PM_{2.5}$, NO_2, benzene and EC for one year at 24 schools located within 400 m of major roadways. While associations were seen be tween symptoms and truck traffic and measured pollutants, there was no significan association between any of the environmental measures and FVC < 85% or FEV1 < 85%. Restricting the analysis to children living within 500 m of highways generall increased ORs.

Personal exposure monitoring of NO_2 as a surrogate for total traffic pollutants wit 298 Korean college students found statistically significant associations with FEV1 FEV1/FVC, and forced expiratory volume between 25 and 75% (FEV25–75), but no with FVC. The multivariate regression model presented suggests that FEV25–75 wa the outcome measure that most clearly showed an effect [75]. Cross-sectional studie of children in Korea [76] and France [77] also indicate that lung function is diminishe in association with area pollutants that largely derive from traffic.

Time series studies suggest there are also acute effects. A study of 19 asthmati children measured PM via personally carried monitors, at homes and at central sit monitors. The study found deficits in FEV1 that were associated with PM, althoug many sources besides traffic contributed to exposure. In addition, the results suggest that ability to see associations with health outcomes improves at finer scale of monitoring [78]. PM was associated with reduced FEV1 and FVC in only the asthmatic subset of children in a Seattle study [79]. Studies have also seen associations between PM and self reported peak flow measurements [80, 81] and asthmatic symptoms [82].

CANCER AND NEAR-HIGHWAY EXPOSURES

As noted above, both the Six-Cities Study [27] and the American Cancer Society (ACS) Study [28] found associations between PM and lung cancer. Follow-up studies using the ACS cohort [29, 37] and the Six-Studies cohort [83] that controlled for smoking and other risk factors also demonstrated significant associations between PM and lung cancer. The original studies were subject to intensive replication, validation, and re-analysis, which confirmed the original findings [84].

The ASHMOG study [85] was designed to look specifically at lung cancer and air pollution among Seventh-day Adventists in California, taking advantage of their low smoking rates. Air pollution was interpolated to centroids of zip codes from ambient air monitoring stations. Highway proximity was not considered. The study found associations with ozone (its primary pollutant of consideration), PM_{10} and SO_2. Notably, these are not the pollutants that would be expected to be substantially elevated immediately adjacent to highways.

A case control study of residents of Stockholm, Sweden modeled traffic-related NO_2 levels at their homes over 30 years and found that the strongest association involved a 20-year latency period [86]. Another case control study drawn from the European Prospective Investigation on Cancer and Nutrition found statistically significantly elevated ORs for lung cancer with proximity to heavy traffic (>10, 000 cars per day) as well as for NO_2 and PM_{10} at nearby ambient monitoring stations [87]. Nafstad et al. [88] used modeled NO_2 and SO_2 concentrations at the homes of over 16,000 men in Oslo to test associations with lung cancer incidence. The models included traffic and point sources. The study found small, but statistically significant associations between NO_2 and lung cancer. Problems that run through all these studies are weak measures of exposure to secondhand tobacco smoke, the use of main roads rather than highways as the exposure group and modeled rather than measured air pollutants.

A study of regional pollution in Japan and a case control study of more localized pollution in a town in Italy also found associations between NO_2 and lung cancer and PM and lung cancer [89, 90]. On the other hand, a study that calculated SIRs for specific cancers across lower and higher traffic intensity found little evidence of an association with a range of cancers [91].

The plausibility of near-highway pollution causing lung cancer is bolstered by the presence of known carcinogens in diesel PM. The US EPA has concluded after reviewing the literature that diesel exhaust is "likely to be carcinogenic to humans by inhalation" [92]. An interesting study of UFP and DNA damage adds credibility to an association with cancer [93]. This study had participants bicycle in traffic in Copenhagen and measured personal exposure to UFP and DNA oxidation and strand breaks in mononuclear blood cells. Bicycling in traffic increased UFP exposure and oxidative damage to DNA, thus demonstrating an association between DNA damage and UFP exposure in vivo.

POLICY AND RESEARCH RECOMMENDATIONS

Based on the literature reviewed above it is plausible that gradients of pollutants next to highways carry elevated health risks that may be larger than the risks of general area ambient pollutants. While the evidence is considerable, it is not overwhelming and is weak in some areas. The strongest evidence comes from studies of development of asthma and reduction of lung function during childhood, while the studies of cardiac health risk require extrapolation from area studies of smaller and larger geographic scales and inference from toxicology laboratory investigations. The lung cancer studies, because they include pollutants such as O_3 that are not locally concentrated, are not particularly strong in terms of the case for near-highway risk. There is a need for lung cancer research that uses major highways rather than heavily trafficked roads as the environmental exposure.

While more studies of asthma and lung function in children are needed to confirm existing findings, especially studies that integrate exposure at school, home and during commuting, to refine our knowledge about the association, we would point to the greater need for studies of cardiac health and lung cancer and their association with near highway exposures as the primary research areas needing to be developed. Many of the studies of PM and cardiac or pulmonary health have focused on mortality. Near highway mortality studies may be possible, but would be lengthy if they were initiated as prospective cohorts. Other possibilities include retrospective case control studies of mortality, cross sectional studies or prospective studies that have end points short of mortality, such as biological markers of disease. For all health end points there is a need for studies that adequately address the possible confounding of SES with proximity to highways. There is good reason to think that property values decline near highways and that control for SES by, for example, income, may be inadequate.

Because of the incomplete development of the science regarding the health risks of near highway exposures and the high cost and implication of at least some possible changes in planning and development, policy decisions are complicated. The State of California has largely prohibited siting of schools within 500 feet of freeways (SB 352; approved by the governor October 2, 2003). Perhaps this is a viable model for other states or for national-level response. As it is the only such law of which we are aware, there may be other approaches that will be and should be tried. One limitation of the California approach is that it does nothing to address the population already exposed at schools currently cited near freeways and does not address residence near freeways.

CONCLUSION

The most susceptible (and overlooked) population in the US subject to serious health effects from air pollution may be those who live very near major regional transportation route, especially highways. Policies that have been technology based and regional in orientation do not efficiently address the very large exposure and health gradients suffered by these populations. This is problematic because even regions that EPA has deemed to be in regional PM "attainment" still include very large numbers of near highway residents who currently are not protected. There is a need for more research, but also a need to begin to explore policy options that would protect the exposed population.

KEYWORDS

- **Black carbon**
- **Particle-bound polycyclic aromatic hydrocarbons**
- **Particulate matter**
- **Socioeconomic status**
- **Ultrafine particulates**

REFERENCES

1. American Housing Survey for the United States: 2003 Series H150/03. Available from: http://www.census.gov/hhes/www/housing/ahs/ahs03/ahs03.html [accessed May 2007].
2. Massachusetts Fact Book; 2004.
3. Chambers LA. Classification and extent of air pollution problems. In: Stern AC, editor. Air Pollution. New York: Academic Press; 1976.
4. Rogge WF, Hildemann LM, Mazurek MA, Cass GR, Simoneit BRT. Sources of fine organic aerosol. 2. Noncatalyst and catalyst-equipped automobiles and heavy-duty diesel trucks. Environmental Science Technology 1993; 27:636–651.
5. Graedel TE, Hawkins DT, Claxton LD. Atmospheric Chemical Compounds: Sources, Occurrence, and Bioassay. New York: Academic Press; 1986.
6. Shi JP, Khan AA, Harrison RM. Measurements of ultrafine particle concentration and size distribution in the urban atmosphere. The Science of the Total Environment 1999; 235:51–64.
7. Zhu Y, Hinds WC, Kim S, Sioutas C. Concentration and size distribution of ultrafine particles near a major highway. Journal of the Air and Waste Management Association 2002; 52:1032–1042.
8. Zhu Y, Hinds WC, Kim S, Shen S, Sioutas C. Study of ultrafine particles near a major highway with heavy-duty diesel traffic. Atmospheric Environment 2002; 36:4323–4335.
9. Zhang KM, Wexler AS, Zhu Y, Hinds WC, Sioutas C. Evolution of particle number distribution near roadways. Part II: the "Road-to-Ambient" process. Atmospheric Environment 2004; 38:6655–6665.
10. Sioutas C, Delfino RJ, Singh M. Exposure assessment for atmospheric ultrafine particles (UFP) and implications in epidemiologic research. Environmental Health Perspectives 2005; 113:947–955.
11. Hitchins J, Morawska L, Wolff R, Gilbert D. Concentrations of submicrometre particles from vehicle emissions near a major road. Atmospheric Environment 2000; 34:51–59.
12. Morawska L, Thomas S, Gilbert D, Greenaway C, Rijnders E. A study of the horizontal and vertical profile of submicrometer particulates in relation to a busy road. Atmospheric Environment 1999; 33:1261–1274.
13. Fischer PH, Hoek G, van Reeuwijk H, Briggs DJ, Lebret E, van Wijnen JH, Kingham S, Elliott PE. Traffic-related differences in outdoor and indoor concentrations of particles and volatile organic compounds in Amsterdam. Atmospheric Environment 2000; 34:3713–3722.
14. Roorda-Knape MC, Janssen NAH, De Hartog JJ, van Vliet PHN, Harssema H, Brunekreef B. Air pollution from traffic in city districts near major motorways. Atmospheric Environment 1998; 32:1921–1930.

15. Janssen NAH, van Vliet PHN, Aarts F, Harssema H, Brunekreef B. Assessment of exposure to traffic related air pollution of children attending schools near motorways. Atmospheric Environment 2001; 35:3875–3884.
16. National Research Council, Committee on Research Priorities for Airborne Particulate Matter. Research priorities for airborne particulate matter, IV: continuing research progress. Washington, D.C.: National Academy Press; 2004.
17. US Environmental Protection Agency. Air quality criteria for particulate matter. Research Triangle Park; 2004.
18. Miller KA, Siscovick DS, Sheppard L, Shepherd K, Sullivan JH, Anderson GL, Kaufman JD. Long-term exposure to air pollution and incidence of cardiovascular events in women. The New England Journal of Medicine 2007; 356:447–458.
19. Riedliker M, Cascio WE, Griggs TR, Herbst MC, Bromberg PA, Neas L, Williams RW, Devlin RB. Particulate matter exposure in cars is associated with cardiovascular effects in healthy young men. American Journal of Respiratory and Critical Care Medicine 2004; 169:934–940.
20. Hoffmann B, Moebus S, Stang A, Beck E, Dragano N, Möhlenkamp S, Schmermund A, Memmesheimer M, Mann K, Erbel R, Jockel KH, Heinz Nixdorf. RECALL Study Investigative Group Residence close to high traffic and prevalence of coronary heart disease. European Heart Journal 2006; 27:2696–2702.
21. Ruckerl R, Greven S, Ljungman P, Aalto P, Antoniades C, Bellander T, Berglind N, Chrysohoou C, Forastiere F, Jacquemin B, von Klot S, Koenig W, Kuchenhoff H, Lanki T, Pekkanen J, Perucci CA, Schneider A, Sunyer J, Peters A. Air pollution and inflammation (IL-6, CRP, fibrinogen) in myocardial infarction survivors. Environmental Health Perspectives 2007; 115:1072–1080.
22. Schwartz J, Litonjua A, Suh H, Verrier M, Zanobetti A, Syring M, Nearing B, Verrier R, Stone P, MacCallum G, Speizer FE, Gold DR. Traffic related pollution and heart rate variability in a panel of elderly subjects. Thorax 2005; 60:455–461.
23. Adar SD, Gold DR, Coull BA, Schwartz J, Stone P, Suh H. Focused exposures to airborne traffic particles and heart rate variability in the elderly. Epidemiology 2007; 18:95–103.
24. Hoek G, Brunekreek B, Goldbohm S, Fischer P, van den Brandt PA. Association between mortality and indicators of traffic-related air pollution in the Netherlands: a cohort study. The Lancet 2002; 360:1203–1209.
25. Peters A, von Klot S, Heier M, Trentinaglia I, Hormann A, Wichmann HE, Lowel H. Exposure to traffic and the onset of myocardial infarction. New England Journal of Medicine 2004; 351:1861–70.
26. Pope CA, Dockery DW. Health effects of fine particulate air pollution: lines that connect. Journal of Air and Waste Management 2006; 56:709–742.
27. Dockery DW, Pope CA, Xu X, Spengler JD, Ware JH, Fay ME, Ferris BG, Speizer FE. An association between air pollution and mortality in six U.S. cities. New England Journal of Medicine 1993; 329:1753–9.
28. Pope CA, Thun MJ, Namboodiri MM, Dockery DW, Evans JS, Speizer FE, Hath CW. Particulate air pollution as a predictor of mortality in a prospective study of US adults. American Journal of Respiratory and Critical Care Medicine. 1995; 151:669–674.
29. Pope CA, Burnett RT, Thun MJ, Calle EE, Krewski D, Ito K, Thurston GD. Lung Cancer, Cardiopulmonary mortality, and long-term exposure to fine particulate air pollution. Journal of the American Medical Association 2002; 287:1132–1141.
30. Kunzli N, Jerrett M, Mack WJ, Beckerman B, LaBree L, Gilliland F, Thomas D, Peters J, Hodis HN. Ambient air pollution and Atherosclerosis in Los Angeles. Environmental Health Perspectives 2005; 113:201–206.
31. Peters A. Particulate matter and heart disease: Evidence from epidemiological studies. Toxicology and Applied Pharmacology 2005:477–482.
32. Wheeler A, Zanobetti A, Gold DR, Schwartz J, Stone P, Suh H. The relationship between ambient air pollution and heart rate variability differs for individuals with heart and pulmonary disease. Environmental Health Perspectives 2006; 114:560–566.

33. Chuang K, Chan C, Chen N, Su T, Lin L. Effects of particle size fractions on reducing heart rate variability in cardiac and hypertensive patients. Environmental Health Perspectives 2005; 113:1693–1697.
34. Chan C, Chuang K, Shiao G, Lin L. Personal exposure to submicrometer particles and heart rate variability in human subjects. Environmental Health Perspectives 2004; 112:1063–1067.
35. Brauer M, Hoek G, van Vliet P, Meliefste K, Fischer P, Gehring U, Heinrich J, Cyrys J, Bellander T, Lewne M, Brunekreef B. Estimating long-term average particulate air pollution concentrations: application of traffic indicators and geographic information systems. Epidemiology 2003; 14:228–239.
36. Brunekreef B, Holgate ST. Air pollution and health. Lancet 2002; 360:1233–1242.
37. Jerrett M, Finkelstein M. Geographies of risk in studies linking chronic air pollution exposure to health outcomes. Journal of Toxicology and Environmental Health 2005; 68:1207–1242.
38. O'Neill MS, Jerrett M, Kawachi I, Levy JI, Cohen AJ, Gouveia N, Wilkinson P, Fletcher T, Cifuentes L, Schwartz J. Workshop on Air Pollution and Socioeconomic Conditions. Health, wealth, and air pollution: advancing theory and methods. Environmental Health Perspectives 2003; 111:1861–1870.
39. Jerrett M, Burnett RT, Ma R, Pope CA, Krewski D, Newbold KB, Thurston G, Shi Y, Finkelstein N, Calle EE, Thun MJ. Spatial analysis of air pollution and mortality in Los Angeles. Epidemiology 2005; 16:727–736.
40. Finkelstein M, Jerrett M, Sears MR. Environmental inequality and circulatory disease mortality gradients. Journal of Epidemiology and Community Health 2005; 59:481–487.
41. Tonne C, Melly S, Mittleman M, Coull B, Goldberg R, Schwartz J. A case-control analysis of exposure to traffic and acute myocardial infarction. Environmental Health Perspectives 2007; 115:53–57.
42. Lipfert FW, Wyzga RE, Baty JD, Miller JP. Traffic density as a surrogate measure of environmental exposures in studies of air pollution health effects: Long-term mortality in a cohort of US veterans. Atmospheric Environment 2006; 40:154–169.
43. Pope CA, Burnett RT, Thurston GD, Thun MJ, Calle EE, Krewski D, Godleski JJ. Cardiovascular mortality and long-term exposure to particulate air pollution—Epidemiological evidence of general pathophysiological pathways of disease. Circulation 2004; 109:71–77.
44. Brook RD, Franklin B, Cascio W, Hong Y, Howard G, Lipsett M, Luepker R, Mittleman M, Samet J, Smith SC, Tager I. Air pollution and cardiovascular disease: a statement for healthcare professionals from the expert panel on population and prevention science of the American Heart Association. Circulation 2004; 109:2655–2671.
45. Sun Q, Wang A, Jin X, Natanzon A, Duquaine D, Brook RD, Aguinaldo JG, Fayad Z, Fuster V, Lippman M, Chen LC, Rajagopalan S. Long-term air pollution exposure and acceleration of atherosclerosis and vascular inflammation in an animal model. Journal of the American Medical Association 2005; 294:3003–3010.
46. Sandhu RS, Petroni DH, George WJ. Ambient particulate matter, C-reactive protein, and coronary artery disease. Inhalation Toxicology 2005; 17:409–413.
47. Oberdorster G. Pulmonary effects of inhaled ultrafine particles. International Archives of Occupational and Environmental Health 2001; 65:1531–1543.
48. Delfino RJ, Sioutas C, Malik S. Potential role of ultrafine particles in associations between airborne particle mass and cardiovascular health. Environmental Health Perspectives 2005; 113:934–946.
49. Venn A, Lewis S, Cooper M, Hubbard R, Hill I, Boddy R, Bell M, Britton J. Local road traffic activity and the prevalence, severity, and persistence of wheeze in school children: combined cross sectional and longitudinal study. Occupational & Environmental Medicine 2000; 57:152–158.
50. Waldron G, Pottle B, Dod J. Asthma and the motorways—One district's experience. Journal of Public Health Medicine 1995; 17:85–89.
51. Lewis SA, Antoniak M, Venn AJ, et al. Secondhand smoke, dietary fruit intake, road traffic exposures, and the prevalence of asthma: A cross-sectional study in young children. American Journal of Epidemiology 2005; 161:406–411.

52. English P, Neutra R, Scalf R, Sullivan M, Waller L, Zhu L. Examining associations between childhood asthma and traffic flow using a geographic information system. Environmental Health Perspectives 1999; 107:761–767.
53. Heinrick J, Topp R, Gerring U, Thefeld W. Traffic at residential address, respiratory health, and atopy in adults; the National German Health Survey 1998. Environmental Research 2005; 98:240–249
54. Van Vliet P, Knape M, de Hartog J, Janssen N, Harssema H, Brunekreef B. Motor vehicle exhaust and chronic respiratory symptoms in children living near freeways. Environmental Research 1997; 74:122–132.
55. Venn AJ, Lewis SA, Cooper M, Hubbard R, Britton J. Living near a main road and the risk of wheezing illness in children. American Journal of Respiratory and Critical Care Medicine 2001; 164:2177–2180.
56. Venn A, Yemaneberhan H, Lewis S, Parry E, Britton J. Proximity of the home to roads and the risk of wheeze in an Ethiopian population. Occupational and Environmental Medicine. 2005; 62:376–380.
57. McConnell R, Berhane K, Yao L, Jerrett M, Lurmann F, Gilliland F, Kunzli N, Gauderman J, Avol E, Thomas D, Peters J. Traffic susceptibility, and childhood asthma. Environmental Health Perspectives 2006; 114:766–772.
58. Nicolai T, Carr D, Weiland SK, Duhme H, von Ehrenstein O, Wagner C, von Mutius E. Urban traffic and pollutant exposure related to respiratory outcomes and atopy in a large sample of children. European Respiratory Journal 2003; 21:956–963.
59. Ryan PH, LeMasters , Biswas P, Levin L, Hu S, Lindsey M, Bernstein DI, Lockey J, Villareal M, Hershey GKH, Grinshpun SA. A comparison of proximity and land use regression traffic exposure models and wheezing in infants. Environmental Health Perspectives 2007; 115:278–284.
60. Kim JJ, Smorodinsky S, Lipsett M, Singer BC, Hodgson AT, Ostro B. Traffic-related air pollution near busy roads: The East Bay children's respiratory health study. American Journal of Respiratory and Critical Care Medicine 2004; 170:520–526.
61. Ong P, Graham M, Houston D. Policy and programmatic importance of spatial alignment of data sources. American Journal of Public Health 2006; 96:499–504.
62. Hwang BF, Lee YL, Lin YC, Jaakkola JJ, Guo YL. Traffic related air pollution as a determinant of asthma among Taiwanese school children. Thorax 2005; 60:467–473.
63. Migliaretti G, Cadum E, Migliore E, et al. Traffic air pollution and hospital admissions for asthma: A case control approach in a Turin (Italy) population. International Archives of Occupational and Environmental Health 2005; 78:164–169.
64. Lweguga-Mukasa JS, Oyana TJ, Johjnson C. Local ecological factors, ultrafine particulate concentrations, and asthma prevalence rates in Buffalo, New York, neighborhoods. Journal of Asthma 2005; 42:337–348.
65. Gauderman WJ, Avol E, Lurmann F, Kuenzli N, Gilliland F, Peters J, McConnell R. Childhood asthma and exposure to traffic and nitrogen dioxide. Epidemiology 2005; 16:737–743.
66. Ryan PH, LeMasters GK, Biswas P, Levin L, Hu S, Lindsey M. A comparison of proximity and land use regression traffic exposure models and wheezing in infants. Environmental Health Perspectives 2007; 115:278–284.
67. Brauer M, Hoek G, Smit HA, de Jongste JC, Gerritsen J, Postma DS, Kerkhof M, Brunekreef B. Air pollution and development of asthma, allergy and infections in a birth cohort. European Respiratory Journal 2007; 29:879–888.
68. Wjst M, Reitmeir P, Dodd S, Wulff A, Nicolai T, von Loeffelholz-Colberg EF, von Mutius E. Road traffic and adverse effects on respiratory health in children. British Medical Journal 1993; 307:596–307.
69. Brunekreef B, Janssen NA, de Hartog J, Harssema H, Knape M, van Vliet P. Air pollution from truck traffic and lung function in children living near motorways. Epidemiology 1997; 8:298–303.
70. Gauderman WJ, McConnell , Gilliland F, London S, Thomas D, Avol E, Vora H, Berhane K, Rappaport EB, Lurmann F, Margolis HG, Peters J. Association between air pollution and lung

function growth in Southern California Children. American Journal of Respiratory and Critical Care Medicine 2000; 162:1383–1390.
71. Gauderman WJ, Avol E, Gilliland F, Vora H, Thomas D, Berhane K, McConnell R, Kuenzli N, Lurmann F, Rappaport E, Margolis H, Bates D, Peters J. The effect of air pollution on lung development from 10 to 18 years of age. New England Journal of Medicine 2005; 351:1057–67.
72. Merkus PJFM. Air pollution and lung function. New England Journal of Medicine 2005; 351:2652.
73. Gauderman WJ, Vora H, McConnell R, Berhane K, Gilliland F, Thomas D, Lurmann F, Avol E, Kunzli N, Jarrett M, Peters J. Effect of exposure to traffic on lung development from 10 to 18 years of age: A cohort study. The Lancet 2007; 369:571–577.
74. Janssen NA-H, Brunekreef B, van Vliet P, Aarts F, Meliefste K, Harssema H, Fischer P. The relationship between air pollution from heavy traffic and allergic sensitization, bronchial hyperresponsiveness, and respiratory symptoms in Dutch school children. Environmental Health Perspectives 2003; 111:1512–1518.
75. Hong Y-C, Leem J-H, Lee K-H, Park D-H, Jang J-Y, Kim S-T, Ha E-H. Exposure to air pollution and pulmonary function in university students. International Archives of Occupational and Environmental Health 2005; 78:132–138.
76. Kim HJ, Lim DH, Kim JK, Jeong SJ, Son BK. Effects of particulate matter (PM10) on pulmonary function of middle school children. Journal of the Korean Medical Society 2005; 20:42–45.
77. Penard-Morand C, Charpin D, Raherison C, Kopferschmitt C, Caillaud D, Lavaud F, Annesi-Maesano I. Long-term exposure to background air pollution related to respiratory and allergic health in schoolchildren. Clinical and Experimental Allergy 2005; 35:1279–1287.
78. Delfino RJ, Quintana PJE, Floro J, Gastanaga VM, Samimi BS, Klienman MT, Liu LJ, Bufalino C, Wu C, McLaren CE. Association of FEV1 in asthmatic children with personal and microenvironment exposure to airborne particulate matter. Environmental Health Perspectives 2004; 112:932–941.
79. Koenig JQ, Larson TV, Hanley QS, Rebolledo V, Dumler K, Checkoway H, Wang SZ, Lin D, Pierson WE. Pulmonary function changes in children associated with fine particulate matter. Environmental Research 1993; 63:26–38.
80. Van der Zee SC, Hoek G, Boezen HM, Schouten JP, van Wijnen JH, Brunekreef B. Acute effects of urban air pollution on respiratory health of children with and without chronic respiratory symptoms. Occupational and Environmental Medicine 1999; 56:802–813.
81. Pekkenen J, Timonen KL, Ruuskanen J, Reponen A, Mirme A. Effects of ultrafine and fine particulates in urban air on peak expiratory flow among children with asthmatic symptoms. Environmental Research 1997; 74:24–33.
82. Peters JM, Avol E, Navidi W, London SJ, Gauderman WJ, Lurmann F, Linn WS, Margolis H, Rappaport E, Gong H, Thomas DC. A study of twelve Southern California communities with differing levels and types of air pollution: Prevalence of respiratory morbidity. American Journal of Respiratory and Critical Care Medicine 1999; 159:760–767.
83. Laden F, Schwartz J, Speizer FE, Dockery DE. Reduction in fine particulate air pollution and mortality: extended follow-up of the Harvard six-cities study. American Journal of Respiratory and Critical Care Medicine 2006; 173:667–672.
84. Health Effects Institute. Reanalysis of the Harvard six cities study and the American Cancer Society study of particulate air pollution mortality. Boston, Mass.; 2000.
85. Beeson WL, Abbey DE, Knutsen SF. Long-term concentrations of ambient air pollutants and incident lung cancer in California adults: Results from the ASHMOG study. Environmental Health Perspectives 1998; 106:813–823.
86. Nyberg F, Gustavsson P, Jarup L, Bellander T, Berglind N, Jakobsson R, Pershagen G. Urban air pollution and lung cancer in Stockholm. Epidemiology 2000; 11:487–495.
87. Vineis P, Hoek G, Krzyzanowski M, Vigna-Tagliani F, Veglia F, Airoldi L, Autrup H, Dunning A, Garte S, Hainaut P, Malaveille C, Matullo G, Overvad K, Raaschou-Nielsen O, Clavel-Chapelon F, Linseisen J, Boeing H, Trichopoulou A, Palli D, Peluso M, Krogh V, Tumino R, Panico S, Bueno-De-Mesquita HB, Peeters PH, Lund EE, Gonzalez CA, Martinez C, Dorronsoro M, Barricarte

A, Cirera L, Quiros JR, Berglund G, Forsberg B, Day NE, Key TJ, Saracci R, Kaaks R, Riboli E. Air pollution and risk of lung cancer in a prospective study in Europe. International Journal of Cancer 2006; 119:169–174.
88. Nafstad P, Haheim LL, Oftedal B, Gram F, Holme I, Hjermann I, Leren P. Lung cancer and air pollution: A 27-year follow up of 16 209 Norwegian men. Thorax 2003; 58:1071–1076.
89. Choi K-S, Inoue S, Shinozaki R. Air pollution, temperature, and regional differences in lung cancer mortality in Japan. Archives of Environmental Health 1997; 52:160.
90. Biggeri A, Barbone F, Lagazio C, Bovenzi M, Stanta G. Air pollution and lung cancer in Trieste, Italy: Spatial analysis of risk as a function of distance from sources. Environmental Health Perspectives 1996; 104:750–754.
91. Visser O, van Wijnen JH, van Leeuwen FE. Residential traffic density and cancer incidence in Amsterdam, 1989–1997. Cancer Causes & Control 2004; 15:331–339.
92. US Environmental Protection Agency. Health Assessment Document for Diesel Engine Exhaust. Washington, D.C.; 2002.
93. Vinzents PS, Meller P, Sorensen M, Knudsen LE, Hertel O, Jensen FP, et al. Personal exposure to ultrafine particulates and oxidative DNA damage. Environmental Health Perspectives 2005; 113:1485–1490.

6 Asthma Triggers in Indoor Air

Theodore A. Myatt, Taeko Minegishi, Joseph G. Allen, and David L. MacIntosh

CONTENTS

INTRODUCTION

Reducing exposure to environmental agents indoors shown to increase asthma symptoms or lead to asthma exacerbations is an important component of a strategy to manage asthma for individuals. Numerous investigations have demonstrated that portable air cleaning devices can reduce concentrations of asthma triggers in indoor air; however, their benefits for breathing problems have not always been reproducible. The potential exposure benefits of whole house high efficiency in-duct air cleaners for sensitive subpopulations have yet to be evaluated.

We used an indoor air quality modeling system (CONTAM) developed by NIST to examine peak and time-integrated concentrations of common asthma triggers present in indoor air over a year as a function of natural ventilation, portable air cleaners, and forced air ventilation equipped with conventional and high efficiency filtration systems. Emission rates for asthma triggers were based on experimental studies published in the scientific literature.

Forced air systems with high efficiency filtration were found to provide the best control of asthma triggers: 30–55% lower cat allergen levels, 90–99% lower risk of respiratory infection through the inhalation route of exposure, 90–98% lower environmental tobacco smoke (ETS) levels, and 50–75% lower fungal spore levels than the other ventilation/filtration systems considered. These results indicate that the use of high efficiency in-duct air cleaners provide an effective means of controlling allergen levels not only in a single room, like a portable air cleaner, but the whole house.

These findings are useful for evaluating potential benefits of high efficiency in-duct filtration systems for controlling exposure to asthma triggers indoors and for the design of trials of environmental interventions intended to evaluate their utility in practice.

BACKGROUND

Asthma is chronic inflammatory disorder of the airways that induces a range of subclinical and clinical effects including but not limited to hyperresponsiveness, airflow limitation, and respiratory symptoms. Approximately 6.7% of adults and 8.5% of children in the United States are reported to suffer from asthma with the greatest prevalence among non-Hispanic black and Hispanic children under 18 years of age [1]. Triggers of asthma exacerbation are varied and include viral infections, certain animal allergens and criteria air pollutants, mites, environmental tobacco smoke (ETS), mold, chemical irritants, and exercise in cold air [2]. Reducing exposure to environmental agents shown to increase asthma symptoms or lead to asthma exacerbations is an important component of a strategy to manage asthma for individuals [3].

Numerous investigations have demonstrated that indoor air cleaning devices can reduce concentrations of asthma triggers in indoor air [4–10]. Some studies have reported associations between use of air cleaners and improvements in respiratory symptoms and breathing problems for children and adults with asthma or persistent allergic rhinitis [11–14]. However, the benefits of air cleaners for breathing problems have not always been reproducible [14–19]. An expert panel recently determined that the evidence offered by health studies is not sufficient to conclude that operation of indoor air cleaning devices alleviates asthma symptoms or improves pulmonary function [14, 18–20].

The heterogeneity in results of air cleaner intervention studies for asthma symptoms may reflect in part the limited efficacy of the portable air cleaners used to mitigate exposure to airborne asthma triggers. Portable air cleaners typically have flow rates of 170–340 cubic meters per hour (m^3/hr) and removal efficiency for fine particle mass ($PM_{2.5}$) of only about 70% because of bypass around their high efficiency particle arrestance (HEPA) filters [21]. For a typical U.S. home size of 450 m^3, a 180 m^3/hr portable device has a theoretical particle removal rate of approximately 0.4 per hour (hr^{-1}), about the same as the air exchange rate for a closed home. Air-flow rates through room filters must be equivalent to several air changes per hour in order to achieve substantial control of airborne particulate matter [4]. In contrast, whole house, high efficiency air cleaning systems that can provide clean air delivery rates up to 10 times greater than a portable air cleaner and particle removal rates of approximately 7 per hour are now available for residences [22]. The mitigation of asthma triggers in indoor air by these systems and potential health benefits for sensitive subpopulations have yet to be evaluated.

To address this knowledge gap, we used an indoor air quality modeling system to examine peak and time-integrated concentrations of fungal spores, environmental tobacco smoke, respiratory viruses, and cat allergen in indoor air associated with natural ventilation, portable air cleaners, and forced air ventilation equipped with conventional and high efficiency filtration systems. As part of the modeling, we simulated several conditions that correspond to asthma management guidance published by the American Lung Association and the National Institutes of Health.

METHODS

We used the CONTAM multi-zone indoor air quality model developed by the National Institute of Standards and Technology (NIST) to estimate indoor concentrations of indoor allergens and irritants associated with asthma [23]. Airflow among indoor and outdoor zones of the building (i.e. rooms and ambient air) in CONTAM occurs via flow paths such as doors, windows, and cracks. Inter-zonal flow is based on the empirical power law relationship between airflow and the pressure difference across a flow path. Simulation of a mechanical ventilation system in CONTAM also induces circulation of air in CONTAM. After airflow among zones is established, mass balance equations are used to calculate pollutant concentrations based on the sources and sinks in each zone. Each zone (i.e. rooms, hallways) is treated as a single node wherein the air has uniform, well-mixed conditions throughout. Performance evaluations of CONTAM have demonstrated that the model simulations of inter-zonal flow and air exchange rate are within 15% on average of corresponding values measured in a single-family home and test home, respectively [24–29].

Our analysis included two residential building templates developed by NIST, a two story detached home and a single story detached home. Single-family detached homes represent over 60% of the total housing stock in the U.S. [30]. The floor areas for the single story and two story-building templates are 180 square meters (m2) and 276 m2 respectively. The templates were based on the U.S. Census Bureau American Housing Survey [31] and the U.S. Department of Energy Residential Energy Consumption Survey [32] and were intended to represent typical U.S. residential building stock [33]. We modified the NIST templates to allow for natural ventilation and leakage through and around windows sized to 11.5% of the area of each wall [34].

Both residential templates were modeled with six different ventilation and filtration configurations (see Table 1). The first configuration was a home with natural ventilation (N) and no capacity for indoor air cleaning. The remaining configurations each employ a central forced air heating and cooling system with differing degrees of filtration including: a standard 1-inch media filter (C), a standard 5-inch media filter (C5), the 1-inch filter with one portable HEPA unit in a bedroom (C+1P), the 1-inch media filter with a portable HEPA unit in the bedroom and one in the living/family room (C+2P), and a high efficiency electrostatic air cleaner with HEPA-like removal efficiency for aerosols (HE).

TABLE 1 Ventilation/Filtration Configuration Information.

Abbreviation	Description
N	Natural ventilation with no air cleaning capacity
Forced Air Systems	
C	Conventional 1-inch media filter (MERV 2)
C5	Standard 5 inch media filter. Based on Perfect Fit 5 inch media filter, Model BAYFTAH26M, Trane Residential Systems, Tyler, TX, USA (MERV 8)
HE	High Efficiency System – CleanEffects™ Model TFD235ALAH000AA, Trane Residential Systems, Tyler, TX, USA
Forced Air Systems plus Portable Air Cleaners	
C+1P	Conventional 1-inch filter plus portable HEPA filter devices. Flow characteristics based on Quiet Flo HEPA Air Purifier Model 20316, Hunter Fan Company, Memphis, TN, USA. Filtration capacity based on Chen et al. (2006).
C+2P	Conventional 1-inch filter plus 2 portable HEPA filters devices (See above)

Homes with central systems were assumed to have air-handling units (AHU) balanced to provide 0.18 $m^3/min/m^2$ (0.6 cfm/ft^2) of air to each room in the house. The duty schedule during heating and cooling periods was simulated with 1-hour resolution based on output from representative runs of the EnergyPlus Energy Simulation Software [35]. In general, the fraction of each hour devoted to forced air heating or cooling was proportional to the difference between ambient temperature and a set point of 22°C (72°F). Hourly duty schedules ranged from 4 minutes per hour during temperate periods to 38 minutes per hour during extreme summer periods and 52 min during extreme winter periods. In simulations with the C1 and C5 filters, a conventional AHU that operated only during periods of heating or cooling demand was used. In the simulations with the high efficiency electrostatic air cleaner, we modeled a modern AHU equipped with a variable speed fan that operates at full speed during periods of heating and cooling demand and at half-speed during all other times. Portable air cleaners were modeled as operating at 118 m^3/hr for 24 hours per day. For the single story home, the return air duct AHU was located in the living room, for the two-story home, there was a return in the hallway of both the first and second story. An air supply diffuser was located in each room of both housing templates.

For simulations of central forced air systems, removal efficiencies for in-duct air cleaners were based on particle size-specific results observed in our prior assessment of in-duct air cleaning technologies conducted in a fully instrumented test home [22]. In that work, we found that the removal efficiency of a polydisperse test dust achieved by in-duct devices (specifically, 1-inch, 5-inch, and high efficiency electrostatic) was approximately 10% lower than the rated efficiencies determined according to ASHRAE Method 52.2, an industry standard performance metric [36]. Through diagnostic testing, we determined that the difference between the rated and in-use performance was the result of bypass where 10% of the airflow through the AHU fan entered the AHU cabinet downstream of the filter bay.

For the portable air cleaners, removal efficiencies were based on studies conducted for the National Center of Energy Management and Building Technologies [21]. Similar to the whole house testing, Chen et al. found approximately 30% leakage in portable units and that none of the portable air cleaners reached HEPA-like filtration.

Meteorological information is used by CONTAM to simulate force convection, radiant leakage, and corresponding air exchange rates. We used year 2005 meteorological data, including hourly wind direction and speed, dry and wet bulb temperature, relative humidity, and cloud cover data, obtained from the National Weather Service for the Cincinnati, Ohio area (Cincinnati/Northern Kentucky International Airport). We chose this area because Cincinnati has four distinct seasons and differences in ventilation are expected to vary by climatic conditions.

Using a temperature-based probabilistic approach based on data from an EPA analysis [37], window and door opening schedules were generated that produced total ventilation rates for centrally and naturally ventilated periods consistent with corresponding air exchange rates determined from field campaigns reported elsewhere [38–40]. During periods in which the windows were open, 40% of the total window area was assumed to be open. The AHU duty schedule and the window schedules were linked so that the AHU was never running when the windows were open. The front door was set to a schedule of opening for 15 minutes five times each day. Particle-size specific deposition rates to indoor surfaces were based on research by Thatcher and colleagues [41]. Due to limitations of the model, deposition rates were assumed independent of air exchange rate and the AHU duty schedule.

A set of indoor allergens and irritants that can play a significant role in triggering asthma attacks was the focus of our analysis. Generation rates and particle size distributions of the contaminants were based on experimental data available in the published literature. Details regarding inputs to the model for the allergens and irritants are presented in Table 2.

Cat Allergen

Emission rates for cat allergen were based on studies that characterized the occurrence, suspension, and removal of cat allergen, Fel d 1, inside homes [7, 42, 43]. Based on findings from those studies, we chose to model generation of cat allergen with a constant and intermittent source. The constant source was used to represent Fel d 1 levels in air during quiescent periods. The intermittent source represented resuspension of cat allergen caused by certain activities such as vacuuming or sitting on a couch [10, 44]. The intermittent source released a burst of allergen once an hour during typical waking hours, 7:00 AM–10:00 PM. The constant generation source was located in

TABLE 2 Model Inputs for Contaminant Emission Rates and Filtration Removal Efficiency Rates.

Contaminant /Particle size	Emission Rate	Deposition Rate (hr^{-1})	1-inch (%)	5-inch (%)	High Efficiency (%)	Portable (%)
Cat Allergen[a]						
0.54	0.0688 μg/hr	0.052	2.5	29.2	90.7	71
0.875	0.0688 μg/hr	0.15	2.5	29.2	90.7	71
1.6	0.1376 μg/hr	0.35	20.7	47	91.8	71
2.7	0.5502 μg/hr	1	20.7	47	91.8	71
4	1.8895 μg/hr	2.2	55.3	77.8	96.5	72
5.25	2.0953 μg/hr	3.5	55.3	77.8	96.5	80
7.4	5.5885 μg/hr	6.5	74.3	86.9	98.4	80
9	10.899 μg/hr	10	74.3	86.9	98.4	80
ETS[b]						
0.0575	1.31 mg/cig	0.02	0	14.6	90.1	70
0.1475	2.84 mg/cig	0.005	0	14.6	90.1	70
0.31	2.84 mg/cig	0.018	0	14.6	90.1	70
0.71	1.31 mg/cig	0.08	2.5	29.2	90.7	71

TABLE 2 *(Continued)*

Contaminant /Particle size	Emission Rate	Deposition Rate (hr^{-1})	1-inch (%)	5-inch (%)	High Efficiency (%)	Portable (%)
Outdoor Fungal Spores						
2.5	NA	0.9	14	47	91.8	71
Virus[c]						
2.1	35.3 q/hr	0.6	14	47	91.8	71
4.5	29.4 q/hr	2.8	55	77.8	96.5	72
7.3	1.8 q/hr	6.5	73	86.9	98.4	80
9.4	0.5 q/hr	10	74	86.9	98.4	80

[a] Between the hours of 7 am – 10 pm, the cat allergen concentration increases for 33% from the intermittent allergen release. Emission rates based on Custovic et al. [76].
[b] A total of 8 cigarettes per day. Per cigarette emission rates (mg/cigarette) based on Klepeis et al. [46].
[c] Emission rate of infectious doses (or quanta) per hour (q/hr) based on Liao et al. [56].

all rooms of the house other than the bedrooms, while the burst source was released only in the main living space (i.e. living room for template 72 and family room for template 28). We omitted release of cat allergen in bedrooms in order to evaluate the extent to which allergen avoidance achieved by restricting cats from bedrooms as recommended by the NIH (2007), may be influenced by the use and efficacy of indoor air cleaning systems.

Aerosols that contain cat allergen range in aerodynamic diameter from less than 0.4 micrometers (μm) to greater than 9 μm [7, 45]. Previous research has demonstrated removal of airborne cat allergen by portable air cleaners with HEPA filters [7]. For the electrostatic air cleaner, we assumed that the removal efficiency of cat allergen was equivalent to the particle-size specific performance observed for standard test dust and described elsewhere [22].

Environmental Tobacco Smoke

Particle-size information and emission rates for ETS were based on information reported from studies of cigarette smoke in experimental chambers [46]. The total particle mass released for each cigarette was 8.3 mg with a release rate of 1.3 mg/min. A recent national survey indicates that the average adult smoker in the United States consumes 15 cigarettes per day [47]. Taking into account waking hours spent at home [48], we modeled ETS emissions as cigarette consumption within the home twice in the morning hours and six times in the evening hours. All cigarettes were assumed to be smoked in the main living space (i.e. living room for template 72 and family room for template 28). Particle size for ETS has been reported to range from 0.05 μm to 0.71 μm [49]. Removal efficiency for ETS is one component of the industry standard method for determining and rating the performance of indoor air cleaning technologies [50].

Outdoor Fungi

In contrast to the other asthma triggers that were modeled as indoor sources, we modeled indoor air concentrations of airborne fungi that result from penetration of mold spores in ambient air. To coincide with the meteorological data noted earlier, daily mold spore counts for February 14 to November 23, 2005 measured at the Hamilton County Environmental Services Office in Cincinnati were obtained from the Hamilton County Air Quality Management Division. The daily observations from Cincinnati are short-term samples collected with a Rotorod Sampler (Sampling Technology, Inc., Minnetonka, MN) and therefore do not reflect the temporal variability of spore concentrations that may occur over the course of each day. In the absence of more complete data, we assumed that concentrations within the day were constant for purposes of this analysis. The outdoor level of total fungal spores reported in the data for Cincinnati ranged between 32 and 7935 spores per cubic meter (spores/m^3) with a geometric mean of 881 spores/m^3. As expected, outdoor spore concentrations were highest in the summer and early fall months. The aerodynamic diameter size distribution for total spores is large, ranging from 1 to 40 μm. While the dominant fungal genera, Cladosporium, has a aerodynamic diameter slightly less than 2 μm [49], the other dominant types, basidiospores and ascospores have aerodynamic diameters on the order of 5 μm [51].

Fungal allergens are borne on spores larger than 2.5 μm as well as hyphael fragments and fragmented spores smaller than 2.5 μm. Because of the absence of information on fungal fragment levels in outdoor spore data for Cincinnati and the paucity of large spore types in the data, we established 2.5 μm as a reasonable central estimate of the aerodynamic diameter for fungi in this analysis.

Respiratory Viruses

We modeled the release of two respiratory viruses, influenza virus and rhinovirus, because they have been implicated as triggers of asthma exacerbations and essential information is available on their transmission [52] and aerosol properties. While respiratory syncytial virus and other viruses have also been associated with asthma, a lack of key information on these organisms precluded their inclusion in this analysis. For our respiratory virus modeling we utilized the concept of infectious dose, referred to as quanta, as first described in 1955 [53] Estimates of quanta generation rates from an infectious person are based on analyses of outbreaks of infectious diseases as described elsewhere [54, 55]. The greater the quanta generation rate the more infectious the organism. Estimates for influenza, a virus that can spread rapidly, are on the order of 15 to 128 quanta per hour [54, 56]. Organisms with slower spreading infections, like rhinovirus and tuberculosis have generation rates on the order of 1 to 10 quanta per hour [54]. For this analysis, we assumed the approximate mid-point of published quanta generation rates for influenza and rhinovirus, 67 q/hr and 5 q/hr, respectively.

We also assumed the quanta were evenly distributed among the particles released during a sneeze. The removal processes are based on the particle sizes of the quanta released. We based the particle size distribution on experimental studies of particles emitted during sneezes and coughs conducted in the 1940s and 1960s [57, 58] and recently reanalyzed [59]. To establish removal efficiency for respiratory virus achieved by the in-duct media filters, we relied upon size-specific results observed in our test home [22]. For the in-duct electrostatic air cleaner, the removal efficiency was based on laboratory studies in which a suspension of live influenza A virus, PR-8 strain (Advanced Biotechnology, Inc., MD) in phosphate buffered saline was aerosolized within a ventilation duct using a 6-jet Collison nebulizer. Aerosol samples were obtained on Teflo filters (Millipore Corporation, Bedford, MA) in triplicate upstream and downstream of the electrostatic air cleaner on three days. The samples were extracted and assayed for influenza by quantitative polymerase chain reaction (qPCR) following procedures described by Van Elden et al. [60]. The average removal efficiency from the tests was greater than 99% with more precise quantitation limited by the sensitivity of the assay. The removal efficiencies obtained from the laboratory studies were coupled with AHU bypass information for use in the model. Details of this novel application of qPCR will be published elsewhere.

Output from the IAQ model for respiratory virus was expressed as quanta per cubic meter (q/m^3) of indoor air. We used a modified Wells–Riley equation [61] to estimate the risk of infection based on the concentration of quanta in the room from the model output coupled with conventional central estimates of exposure duration and a breathing rate of 0.48 m^3/hr published in a widely used compilation of exposure factors [62]. We used the results to analyze the risk of infection for an individual when

(1) spending time in the same room as an infectious individual, (2) spending time in an adjacent room, and (3) occupying other rooms in the house when an infected individual is either in a bedroom or in the living room of the home.

RESULTS

Air exchange rate (AER) is an influential determinant of indoor air quality and hence is a primary output from the CONTAM model. The distributions of 24-hour average AER across the year for the two templates with both natural and forced air ventilation systems are summarized in Table 3. The mean and median AER for the natural ventilation configuration were approximately twice those in the forced air configuration due to the increased use of windows during warm weather. AER was lower in the newer home (DH28) than the older home (DH72), which reflects differences in leakage rates between the two homes. With the exception of differences in AER, the modeling results were similar for the two home templates. Therefore, we chose to report only the results from the newer two-story home (DH28).

TABLE 3 Distribution of Simulated 24-Hour Average Air Exchange Rates for Homes with and Without Forced Air Ventilation Systems.

Ventilation/Filtration	House Template	Mean	Std Dev	Percentiles				
				5%	25%	50%	75%	95%
Natural	DH28	3.7	5.0	0.1	0.2	0.2	6.8	13.0
Forced Air		1.8	3.6	0.1	0.1	0.2	0.9	10.9
Natural	DH72	3.0	3.9	0.2	0.4	0.5	5.1	10.6
Forced Air		1.6	2.9	0.1	0.3	0.4	0.7	8.7

DH28 Two-story detached home
DH72 Single-story detached home

Cat Allergen

The distribution of hourly average concentrations for airborne cat allergen throughout the home for each of the six ventilation configurations is summarized in Figure 1A. When operating a high efficiency device, the median allergen concentration (4.0 ng/m^3) was 46% lower when compared to conventional filtration (6.4 ng/m^3). The next best performance was achieved by two systems—the in-duct 5-inch media filter (C5) and a portable air cleaner in the same room as the intermittent release of allergen (i.e. C+2P). Nominally, peak concentrations were best mitigated by the high efficiency in-duct device (86 ng/m^3), although the difference in comparison to peaks associated with the other air cleaning approaches (approximately 100 ng/m^3) may not be substantive relative to uncertainties in the modeling analysis. To evaluate the effectiveness of the ventilation configurations at limiting transfer of allergen to bedrooms, all airborne releases of cat allergen in our model occurred outside of the bedrooms. In the bedroom, allergen levels were lower than the whole house average for all configurations, with the high efficiency in-duct filtration performing best at minimizing the transfer of allergen into the bedroom (see Figure 1B).

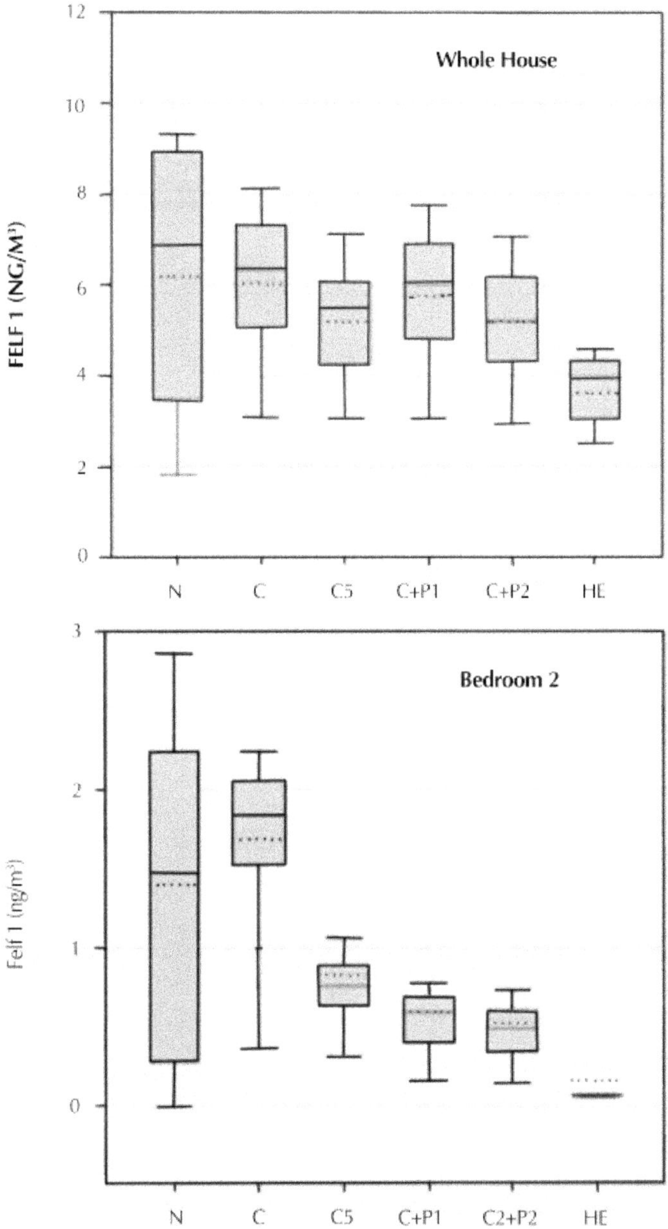

FIGURE 1 Comparison of Hourly Fel d 1 allergen concentrations by filtration configuration for (1A) the whole house average and (1B) bedroom 2.

Environmental Tobacco Smoke

Modeled whole house concentrations of ETS were strongly influenced by use of air cleaners as illustrated by the distribution of hourly average concentrations estimated across the year (Figure 2). The greatest mitigation of ETS was achieved by the high efficiency in-duct device (median <0.01 μg/m³), followed by use of a portable air cleaner in the same room as the smoker (median 3.2 μg/m³), the pleated in-duct media filter (median 9.8 μg/m³), one portable air cleaner in a bedroom (median 17.8 μg/m³), and a conventional in-duct filter (median 29.9 μg/m³). Simulation of a home with natural ventilation yielded hourly average ETS concentrations that were similar to the C5 simulation, probably because of the higher AER throughout the year for a home without forced air conditioning.

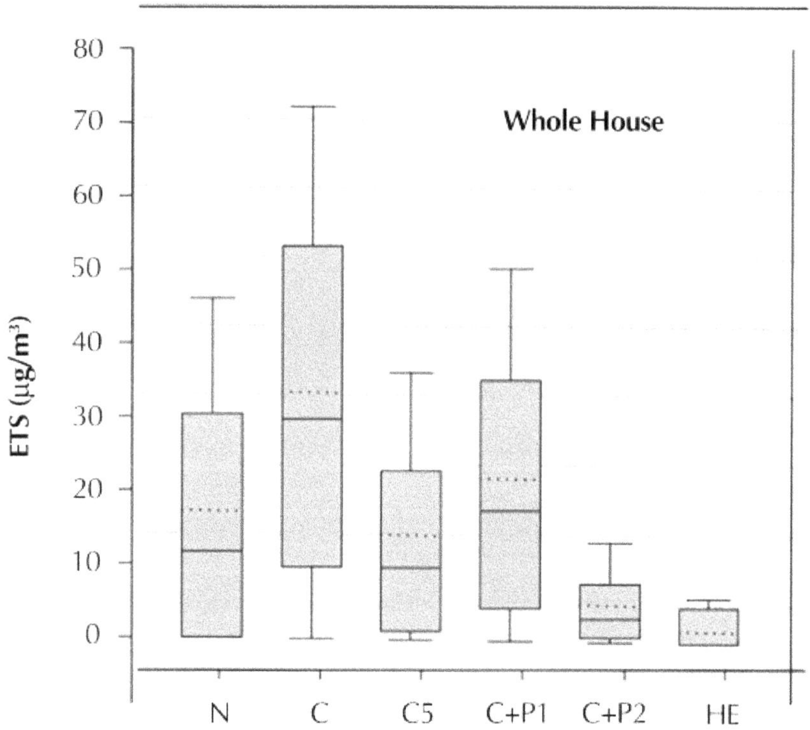

FIGURE 2 Comparison of Hourly ETS concentrations by filtration configuration.

The effect of high efficiency in-duct filtration on peak and short-term time-weighted averaged levels of ETS is depicted in Figure 3A and 3B for a typical 24-hour period (February 1) that had eight smoking events in the living room. For a home with conventional in-duct filtration, each cigarette smoked is associated with a peak concentration of approximately 80 μg/m³ and a subsequent exposure period of at least 8 hours when windows are closed. In contrast, peak ETS concentrations per cigarette during model runs with the high efficiency in-duct device were about 40 μg/m³. First-order removal rates for ETS calculated for the conventional and high efficiency in-duct filtration conditions were 0.008 min^{-1} and 0.049 min^{-1}. Use of the high efficiency in-duct device also substantially limited the distribution of the contaminant into other rooms of the home such as the bedroom.

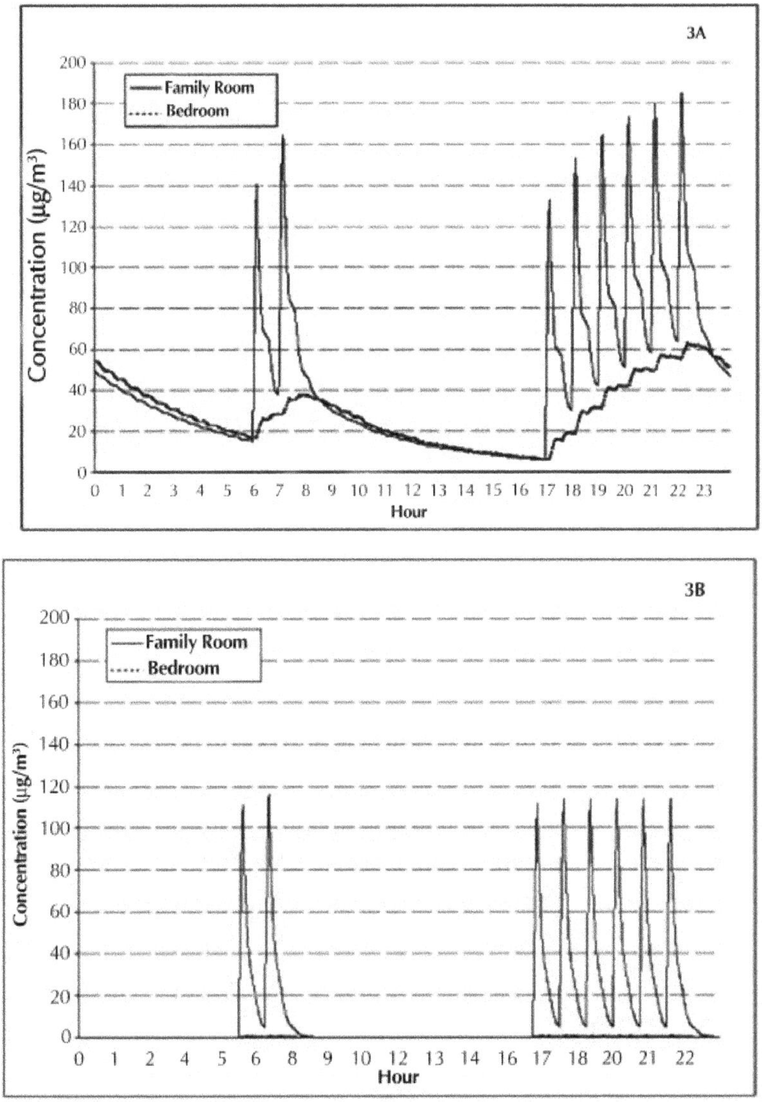

FIGURE 3 Comparison of 24-hour environmental tobacco smoke (ETS) concentrations in the living room and bedroom between the conventional filter (3A) and the high-efficiency filter (3B) for February 1.

Outdoor Fungi

The highest indoor/outdoor ratios for spore concentrations were in the summer and fall months, probably due to the higher AER associated with open windows during those seasons. When averaged over the period for which fungal spore data were available, the indoor/outdoor ratio was highest for the natural ventilation configuration and lowest for the in-duct high efficiency configuration (Table 4). Whole house indoor spore concentrations for the in-duct high efficiency configuration were less than one-half the levels in the conventional configuration and more than eight times lower than the mean outdoor level. Even in the bedroom where the portable air cleaner was located, the in-duct high efficiency achieved lower spore levels.

TABLE 4 Geometric Mean (Geometric Standard Deviation) of Indoor/Outdoor Ratios and Indoor Spore Concentrations by Ventilation/Filtration Type.

Ventilation on/Filtration	I/O Ratio	Whole House (spores/m^3)	Bedroom 2[a] (spores/m^3)
N	0.34 (2.6)	303 (7.0)	238 (9.4)
C	0.16 (2.7)	141 (5.8)	131 (6.3)
C5	0.13 (3.1)	111 (6.7)	97 (8.0)
C+1P	0.14 (2.9)	128 (6.0)	54 (8.5)
C+2P	0.14 (2.9)	119 (6.2)	52 (8.8)
HE	0.07 (4.1)	57 (8.3)	41 (13.0)

[a]Bedroom 2 contains a portable air cleaner

Respiratory Viruses

For the virus results, we limited the analysis to December through March to reflect the cold and flu season in the United States. The median AER for this period was 25% lower for the naturally ventilated configuration and essentially unchanged for the mechanically ventilated homes in comparison to the remainder of the year. For this period, we examined the extent to which the risk of infection by either influenza or rhinovirus is modified by the use of an air cleaner for three common scenarios where a healthy individual and infectious individual cohabitate.

In the first scenario, a healthy individual, perhaps a caregiver, spends one hour in the bedroom of an individual infected with influenza. In this case, the use of a portable air cleaner in the room with the infectious individual limits the average risk of infection to less than one-half the risk when conventional filtration is used (Table 5). The high efficiency in-duct system provides the next lowest average risk of infection, followed by the conventional and pleated filter in-duct systems. The risk of infection is lowered for each of the in-duct and portable air cleaner configurations in comparison to natural ventilation.

TABLE 5 Mean (Standard Deviation) Percent Risk of Infection During Three Exposure Scenarios.

Scenario	1	2	3	
Ventilation /Filtration	Risk of influenza infection for a one hour exposure in the bedroom with individual infected with influenza	Risk of influenza infection from 12 hour exposure in adjacent bedroom	Risk of infection during 5 day infectious period while infected individual in bedroom for 1/2 the day and the family room for the 1/2 the day[a]	
			Influenza	Rhinovirus
N	36 (7.9)	0.6 (1.3)	17.1 (2.4)	1.4 (0.2)
C	18 (3.4)	16.1 (1.7)	70.0 (1.6)	8.6 (0.4)
C5	17 (3.4)	6.7 (1.0)	36.6 (1.8)	3.4 (0.2)
C+1P	7 (0.8)	5.9 (0.7)	51.9 (1.7)	5.3 (0.2)
C+2P	7 (0.8)	5.4 (0.6)	33.7 (2.2)	3.0 (0.2)
HE	13 (1.5)	0.6 (0.1)	3.9 (0.2)	0.3 (0.01)

[a] Assumes that that occupant is in the home 68.7% of the time based on Klepeis et al. [48]

In the second scenario, we evaluated the risk of infection for a person who spends 12 hours in a bedroom adjacent to a second bedroom occupied by an individual infected with influenza. This scenario is representative of many residential configurations including children who normally sleep in separate bedrooms or two children who normally share a bedroom but are separated temporarily when one of them has a chest cold. In this scenario, the risk of influenza infection for a 12-hour exposure for an occupant in the adjacent bedroom was approximately 16% with conventional filtration, 5% for the configurations with a portable air cleaner in the bedroom and 0.6% with the high efficiency filtration (See Table 5).

For the third scenario, we estimated the risk of infection from an individual who remains in the home over the course of a five-day infectious period. We assumed that the infectious individual spent one-half of their time in the bedroom and the other half in the family room, while a healthy individual spent 69% of the corresponding time indoors at home [48] during which they were exposed to the house-wide average concentration of quanta in air. For this scenario, the risk of infection by influenza was greater than 30% in the ventilation configuration with a portable air cleaner in both of the two rooms frequented by the infectious individual (Table 5). In comparison, the risk of infection was 17% for the natural ventilation configuration and less than 4% for the high efficiency in-duct system. The former probably reflects a relatively slow rate of inter-zonal transfer and the latter reflects the comparatively high flow rate and removal efficiency of the in-duct system.

DISCUSSION

Several studies have assessed the use of air cleaners for reducing indoor air concentrations of chemical and biological materials that exacerbate asthma. In these studies, the air cleaning intervention was typically a portable air cleaner sized for a single room of typical size in a residence. Although based on modeling rather than measurements, our analysis indicates that certain air cleaning configurations can mitigate indoor air concentrations of some common asthma triggers more effectively on average than air cleaning achieved by the type of portable filtration devices evaluated previously as well as by conventional in-duct filtration.

Prior performance evaluations of CONTAM demonstrate that the model provides a reasonable degree of accuracy for the types of indoor air quality simulations upon which our analyses rely. Inter-zonal airflow predictions from CONTAM simulations of a single story home were within 15% of corresponding measured values [29]. Similarly, air exchange rates for a single room building predicted with CONTAM were within 5% of measured levels [24]. In a related analysis, the correlation between predicted and observed concentrations of a conservative gas ranged from 0.95 to 0.998 during six tests within a single room test home [25]. In a tracer gas study conducted in a multi-room occupied townhouse, gas concentrations predicted by the model were within 25% of measured concentrations [26]. Finally, measured and predicted 24-hour average concentrations of 0.3 to 5 μm particles in a single room building were within 30% of each other [24].

Particle removal efficiencies for air cleaning systems considered in this analysis were derived from empirical data obtained from test homes or test chambers [21, 22].

Removal efficiencies for the portable air cleaners were based on chamber studies of four different devices that all claimed to have HEPA filters but whose efficacy under controlled conditions was low compared to HEPA standards [21]. If we had assumed that the portable air cleaners had removal efficiencies approaching those of HEPA filters, those systems would have compared more favorably to the other devices for the rooms of the homes in which they were located. Whole house comparisons of portable and in-duct systems are unlikely to have been changed substantially if we had assumed a higher aerosol removal efficiency for the portable devices.

In terms of controlling residence-wide concentrations of cat allergen, ETS, respiratory viruses, and mold spores in indoor air, use of a high efficiency in-duct air cleaner as part of a forced air ventilation system yielded the greatest benefit, followed by multiple portable air cleaners in conjunction with conventional in-duct filtration. The greatest benefit of air cleaning systems over conventional in-duct filtration was observed for ETS, probably because of its sub-micron size distribution and the correspondingly low rate of deposition to surfaces. The extent to which these findings can be generalized to other constituents of indoor air depends upon their similarity in terms of emission profiles and aerodynamic characteristics. Other important indoor allergens such as dust mite and cockroach that have been shown to be associated with relatively particles are unlikely to be represented accurately by our results for cat allergen, ETS, viruses, and fungal spores.

Consistent with results from our evaluation of air cleaners in a test home [22], the whole house performance of each system was directly related to its clean air delivery rate (CADR), the product of airflow rate and removal efficiency. This analysis focused on single-family detached homes, however we anticipate that the findings are applicable to multi-family and attached homes as well. Various types of housing stock may differ systematically in terms of air exchange rate because of variation in construction practices, exterior surface area-to-volume ratios, and other factors. Particle deposition has been reported to be positively associated with air exchange rate due to increased turbulence of indoor air [63, 64]. Because of modeling constraints, we assumed that particle deposition rates were independent of air exchange rate. This simplifying assumption is unlikely to be a substantial contributor to uncertainty in our results because the range of turbulence-induced deposition rates reported for respirable-sized aerosols is small in comparison to differences in performance among air cleaning devices indicated by our analysis.

In terms of controlling the contaminant concentrations in a single room, the location of the contaminant source is important. If the contaminant source was in the family room of the home and therefore near a central return, as was the case for the allergen and ETS modeling, the high efficiency in-duct filtration was superior to all configurations including those with a portable air cleaner in the room. Similarly, if the source is outdoors, as was the case with the fungal spore modeling, the high efficiency in-duct filtration was superior. Conversely, when the source was in a location away from a central return, like a bedroom, as was the case for the one-hour influenza scenario, operation of a portable air cleaner in the room was the most effective air-cleaning configuration. We anticipate that these results for cat allergen, ETS, and virus are reasonably representative of emissions of other respirable-sized aerosols from indoor sources including fungal spores that may be released from surfaces as a result of mechanical forces.

The utility of the modeling results presented is related primarily to relative differences between the air cleaning systems included in this assessment. If reasonable however, the absolute levels are also of interest for consideration of potential air quality and exposure benefits afforded by indoor air cleaning systems. To assess the accuracy of the model results, we compared the predicted concentrations to measurements from studies that quantified residential airborne levels of animal allergens [5, 7, 10], ETS [4, 65], or fungal spores [6, 66]. Several of the studies evaluated the effectiveness of portable air cleaners with HEPA filters which allows us to compare our modeled results to not only the reported levels, but also to the changes in contaminant concentrations associated with use of portable air cleaners. Other studies were designed to measure typical residential contaminant concentrations, both with and without a source present. Data from those investigations provide a reasonable benchmark for our modeled results under typical ventilation configurations.

The relative differences among the ventilation configurations that we considered are similar to the reductions observed in intervention studies designed to evaluate the effectiveness of portable air cleaners. In a study of dog allergen, airborne levels in two rooms with portable air cleaners were reduced to 25% of the baseline allergen level [5]; similar to the difference in modeled cat allergen concentrations for the bedroom when the portable air cleaner was introduced. In a study of portable air cleaner efficacy in four homes with smokers, PM concentrations in the living room were reduced by 30–70% with the use of a portable air cleaner [4]. Our modeling yielded similar reductions in ETS when comparing the conventional filtration to the ventilation configurations with portable air cleaners. In a study designed to evaluate the utility of portable air cleaners for controlling fungal spore concentrations, the intervention effectively reduced spore levels in a bedroom of a residence, however the air cleaner worked best when the bedroom door was closed [6].

When comparing absolute levels of contaminants in the home, the modeled results for the conventional and natural ventilation configurations compare well with values reported in the literature. Our modeled cat allergen concentrations with conventional filtration are similar to concentrations reported in a study of 75 homes with cats in Britain [7] and the levels during our intermittent release of allergen is similar to measured values during periods of disruptions such as vacuuming [10, 67]. In a study of homes in six U.S. cities, the authors calculated that smoking one pack of cigarettes daily contributed 20 $\mu g/m^3$ to the 24 average hour indoor particle concentration [65], which is similar to our modeled ETS concentrations for the conventional and natural ventilation configurations. For fungal spores, our modeled indoor/outdoor ratios for the conventional and natural ventilation configurations are similar to the ratio of 0.32 for total spores reported in a study conducted in six homes in the Cincinnati area [66].

Results from a controlled study of rhinovirus transmission provide a reasonable comparison for evaluating the accuracy of our modeled likelihood of infection. In the experimental study, groups of eight students with active rhinovirus infection spent 12-hours in a room with 12 susceptible students and followed a protocol designed to allow transmission of an infectious dose only by inhalation [52]. The resulting risk of infection from this study was 61%. While AER or filtration characteristics were not reported for this study, we assumed that the room was either naturally ventilated or had conventional filtration. Our modeled scenario with one infectious individual in a room of approximately one-half the size of the experimental room resulted in an average risk of infection with influenza of 33.6% and 16.5% with natural and conventional filtration, respectively. If the modeling were conducted with four infectious individuals in the smaller bedroom to more closely mimic the conditions of the experimental study, the risk of modeled infection would rise to a level similar to that observed in the experimental study.

We relied upon the concept of quanta generation to estimate the probability of acquiring an infection through the airborne route, using the Wells–Riley equation [61]. The Wells–Riley equation and modifications of the equation have been used by researchers to estimate the risk of airborne transmission of an infection for a variety of organisms including measles [61], influenza [54], rhinovirus [54, 68], severe acute respiratory syndrome (SARS) [69], and tuberculosis [55]. The Wells–Riley equation only estimates the risk of transmission for the inhalation route of exposure. Organisms like rhinovirus and influenza can be transmitted by other routes of exposure such as direct contact, although the relative importance of the respective routes of exposure is not well understood. The ability of various indoor air cleaning configurations to influence virus transmission through surface-mediated pathways remains to be determined. Consideration of virus transmission via surfaces and other pathways is unlikely to influence our findings for modification of the risk of infection through inhalation because of different ventilation and air cleaning configurations.

While the Wells–Riley equation accounts for the ventilation rate of the indoor space of interest to calculate the quanta concentration, it does not, as discussed recently [70], explicitly account for other removal processes such as deposition to surfaces, filtration, and loss of infectiousness in the air. However, quanta generation rates are typically based on disease outbreak data, and therefore inherently account for these processes. Our modeling accounted for deposition and filtration, but not loss of infectiousness. Some data suggest that virus die-off is a slow process that can occur over several days at temperature and humidity levels typical of indoor environments [71]. Therefore, not explicitly controlling die-off is unlikely to influence our results substantively. Including removal mechanisms in our model along with the estimates of virus emissions in units of quanta may have resulted in double counting for removal by filtration and deposition. Therefore, our results may underestimate the actual risk of infection. To evaluate the impact of potential double counting for deposition, we conducted model runs without a deposition rate for virus. In these models, the risk of infection increased approximately 30 to 50% depending on the filtration type. Regardless, our analysis was designed to primarily evaluate the differences in ventilation configurations and the differences between these configurations would not be changed by under or over estimating the risk of infection.

While a number of intervention studies clearly demonstrate exposure reductions attributable to the use of portable air cleaners, associated improvements in health have been more difficult to demonstrate. Some air cleaning interventions have yielded improvements in respiratory symptoms and breathing problems for children and adults with asthma or persistent allergic rhinitis [11–13, 72]; however, the results of these studies have not always been reproducible [14–18, 73]. One explanation for the lack of reproducible results could be that portable air cleaners used in these studies have not effectively reduced personal exposure. Our modeling demonstrates that while the use of a portable air cleaner will provide exposure benefits in the room it is located, concentrations of common asthma triggers throughout the residence, and corresponding personal exposures, are not likely to be mitigated. Our modeling analysis indicate that high efficiency in-duct air cleaning systems would yield a more substantial reduction in personal exposure that the portable air cleaners used in intervention studies published to date. Potential benefits of these systems for personal exposure could be evaluated following methods employed in a study of personal exposure to cat allergen [74].

An Expert Panel convened by the NIH recommended asthmatics with pet allergies that are not willing to part with their pets keep the pet out of the asthmatic's bedroom as one part of an asthma management strategy. Additionally, the National Environmental Education & Training Foundation (NEETF) recommends that the use of portable air cleaners in bedrooms of asthmatics [75]. While the use of portable air cleaners in the bedroom prove to be beneficial in our modeling, the results indicate that the use of high efficiency in-duct air cleaners provide an more effective means of controlling allergen levels not only in a single room, but the whole house.

CONCLUSION

The modeling results from this study demonstrate that properly maintained forced air systems with a high efficiency aerosol removal system are expected to provide the best control of the indoor exposure to common asthma triggers such as cat allergen, ETS, fungal spores and respiratory viruses. The modeling results also showed that the potential efficacy of avoidance strategies recommended for asthmatics by the American Lung Association and the National Institutes of Health may be enhanced by the use of certain indoor air cleaning systems.

KEYWORDS

- **Absolute humidity**
- **Indoor moisture levels**
- **Meteorological information**
- **Multi-zone indoor air quality model**
- **Relative humidity**

REFERENCES

1. American Lung Association. State of lung disease in diverse communities: 2007. New York, NY: American Lung Association; 2007. Available at: http://www.lungusa.org/atf/cf/%7b7A8D42C2-FCCA-4604-8ADE-7F5D5E762256%7d/SOLDDC_2007.PDF.

2. National Heart Lung and Blood Institute. Guidelines for the diagnosis and management of asthma (EPR-3). Bethesda, MD: National Heart Lung and Blood Institute; 2007. Available at: http://www.nhlbi.nih.gov/guidelines/asthma/index.htm.
3. Wu F, Takaro TK. Childhood asthma and environmental interventions. Environmental Health Perspectives 2007; 115:971–975.
4. Batterman S, Godwin C, Jia C. Long duration tests of room air filters in cigarette smokers' homes. Environmental Science & Technology 2005; 39:7260–7268.
5. Green R, Simpson A, Custovic A, Faragher B, Chapman M, Woodcock A. The effect of air filtration on airborne dog allergen. Allergy 1999; 54:484–488.
6. Cheng Y, Lu J, Chen T. Efficiency of a portable indoor air cleaner in removing pollens and fungal spores. Aerosol Science and Technology 1998; 29:92–101.
7. Custovic A, Simpson A, Pahdi H, Green RM, Chapman MD, Woodcock A. Distribution, aerodynamic characteristics, and removal of the major cat allergen Fel d 1 in British homes. Thorax 1998; 53:33–38.
8. Abraham ME. Microanalysis of indoor aerosols and the impact of a compact high-efficiency particulate air (HEPA) filter system. Indoor Air 1999; 9:33–40.
9. Hacker DW, Sparrow EM. Use of air-cleaning devices to create airborne particle-free spaces intended to alleviate allergic rhinitis and asthma during sleep. Indoor Air 2005; 15:420–431.
10. de Blay F, Chapman MD, Platts-Mills TA. Airborne cat allergen (Fel d I). Environmental control with the cat in situ. The American Review of Respiratory Disease 1991; 143:1334–1339.
11. Heide S van der, Kauffman HF, Dubois AE, de Monchy JG. Allergen reduction measures in houses of allergic asthmatic patients: effects of air-cleaners and allergen-impermeable mattress covers. European Respiratory Journal 1997; 10:1217–1223.
12. Heide S van der, van Aalderen WM, Kauffman HF, Dubois AE, de Monchy JG. Clinical effects of air cleaners in homes of asthmatic children sensitized to pet allergens. Journal of Allergy and Clinical Immunology 1999; 104:447–451.
13. Francis H, Fletcher G, Anthony C, Pickering C, Oldham L, Hadley E, Custovic A, Niven R. Clinical effects of air filters in homes of asthmatic adults sensitized and exposed to pet allergens. Clinical and Experimental Allergy 2003; 33:101–105.
14. Reisman RE. Do air cleaners make a difference in treating allergic disease in homes? Annals of Allergy, Asthma, and Immunology 2001; 87:41–43.
15. Verrall B, Muir DC, Wilson WM, Milner R, Johnston M, Dolovich J. Laminar flow air cleaner bed attachment: a controlled trial. Annals of Allergy 1988; 61:117–122.
16. Wood RA, Johnson EF, Van Natta ML, Chen PH, Eggleston PA. A placebo-controlled trial of a HEPA air cleaner in the treatment of cat allergy. American Journal of Respiratory and Critical Care Medicine 1998; 158:115–120.
17. Warburton CJ, Niven RM, Pickering CA, Fletcher AM, Hepworth J, Francis HC. Domiciliary air filtration units, symptoms and lung function in atopic asthmatics. Respiratory Medicine 1994; 88:771–776.
18. McDonald E, Cook D, Newman T, Griffith L, Cox G, Guyatt G. Effect of air filtration systems on asthma: a systematic review of randomized trials. Chest 2002; 122:1535–1542.
19. Nelson HS, Hirsch SR, Ohman JL, Platts-Mills TAE, Reed CE, Solomon WR. Recommendations for the use of residential air-cleaning devices in the treatment of allergic respiratory diseases. Journal of Allergy and Clinical Immunology 1988; 82:661–669.
20. National Heart, Lung, and Blood Institue. Expert Panel Report 3: Guidelines for the Diagnosis and Management of Asthma. Bethesda, MD: National Heart, Lung, and Blood Institute; 2007.
21. Chen W, Gao Z, Zhang J, Kosar D, Walker C, Novosel D. Reduced energy use through reduced indoor contamination in residential settings. Alexandria, VA: National Center for Energy Management and Building Technologies; 2006.
22. MacIntosh DL, Myatt T, Ludwig J, Baker BJ, Suh H, Spenger JD. Whole house particle removal and clean air delivery rates for in-duct and portable ventilation systems. Journal of the Air and Waste Management Association 2008; accepted for publication.

23. Walton G, Dols WS. CONTAM 2.1 Supplemental user guide and program documentation. Gaithersberg, MD: National Institute of Standards and Technology; 2003.
24. Emmerich S, Nabinger S. Measurement and simulation of the IAQ impact of particle air cleaners in a single-zone building. Gaithersberg, MD: National Institute of Standards and Technology; 2000. Available at: http://www.bfrl.nist.gov/IAQanalysis/docs/NISTIR6461.pdf.
25. Howard-Reed C, Nabinger S, Emmerich SJ. Characterizing gaseous air cleaner performance in the field. Building and Environment 2008; 43:368–377.
26. Emmerich SJ, Nabinger S, Gupte A, Howard-Reed C, Wallace L. Comparison of measured and predicted tracer gas concentrations in a townhouse. Gaithersburg, MD: National Institute of Standards and Technology; 2003. Available at: http://www.bfrl.nist.gov/IAQanalysis/docs/NISTIR_7035_final11-CC.pdf.
27. Lansari A, Streicher J, Huber A, Crescenti G, Zweidinger R, Duncan J, Weisel C, Burton R. Dispersion of automotive alternative fuel vapors within a residence and its attached garage. Indoor Air 1996; 6:118–126.
28. Howard-Reed C, Nabinger S, Emmerich SJ. Predicting the performance of non-industrial gaseous air cleaners: measurements and model simulations from a pilot study. Gaithersburg, MD: National Institute of Standards and Technology; 2004.
29. Haghighat F, Rao J. A comprehesive validation of two airflow models—COMIS and CONTAM. Indoor Air 1996; 6:278–288.
30. Year 2000 Census. Available at: http://factfinder.census.gov/home/saff/main.html?_lang=en.
31. HUD. The American housing survey for the United States. Washington, DC: U.S. Department of Housing and Urban Development, U.S. Department of Commerce; 1999.
32. U.S. Department of Energy. A look at residential energy consumption in 1997. Washington, DC: U.S. Department of Energy; 1999.
33. Persily A, Musser A, Leber D. A collection of homes to represent U.S. housing stocks. Gaithersberg, MD: National Institute of Standards and Technology; 2006.
34. Enermodal Engineering Limited Characterization of Framing Factors for Low-Rise Residential Building Envelopes. Final report prepared for ASHRAE, Atlanta, GA (USA); 2001. Available at: http://www.energy.ca.gov/title24/2005standards/archive/documents/2001-11-14_workshop/2001-11-07_FRAMING_FACTORS.PDF.
35. U.S. Department of Energy. EnergyPlus Version 2.1.0; 2007.
36. ASHRAE Standard 52.2—1999. Method of testing general ventilation air-cleaning devices for removal efficiency by particle size. Atlanta, GA: American Society of Heating, Refrigeration, and Air Conditioning Engineers; 1999.
37. Johnson T. A guide to selected algorithms, distributions, and databases used in exposure models developed by the office of air quality planning and standards. Research Triangle Park, NC: U.S. Environmental Protection Agency, Office of Research and Development; 2002. Available at: http://www.epa.gov/ttn/fera/data/human/report052202.pdf
38. Sarnat SE, Coull BA, Schwartz J, Gold DR, Suh HH. Factors affecting the association between ambient concentrations and personal exposures to particles and gases. Environmental Health Perspectives 2006; 114:649–654.
39. Suh H, Spenger JD, Koutrakis P. Personal exposures to acid aerosols and ammonia. Environmental Science and Technology 1992; 26:2507–2517.
40. Murray DM, Burmaster DE. Residential air exchange rates in the United States: Empirical and estimated parametric distributions by season and climate region. Risk Analysis 1995; 15:459–465.
41. Thatcher TL, Lai ACK, Moreno-Jackson M, Sextro RG, Nazaroff WW. Effects of room furnishings and air speed on particle deposition rates indoors. Atmospheric Environment 2002; 36.
42. Custovic A, Simpson B, Simpson A, Hallam C, Craven M, Woodcock A. Relationship between mite, cat, and dog allergens in reservoir dust and ambient air. Allergy 1999; 54:612–616.
43. de Blay F, Spirlet F, Gries P, Casel S, Ott M, Pauli G. Effects of various vacuum cleaners on the airborne content of major cat allergen (Fel d 1). Allergy 1998; 53:411–414.

44. Montoya LD, Hildemann LM. Evolution of the mass distribution of resuspended cat allergen (Fel d 1) indoors following a disturbance. Atmospheric Environment 2001; 35:859–866.
45. Green R, Simpson A, Custovic A, Faragher B, Chapman M, Woodcock A. The effect of air filtration on airborne dog allergen. Allergy 1999; 54:484–488.
46. Klepeis NE, Apte MG, Gundel LA, Sextro RG, Nazaroff WW. Determining size-specific emission factors for environmental tobacco smoke particles. Aerosol Science and Technology 2003; 37:780–790.
47. U.S. Department of Health and Human Services. The NSDUH report: Quantity and frequency of cigarette use. Rockville, MD: Office of Applied Studies, U.S. Department of Health and Human Services. Available at: http://www.oas.samhsa.gov/2k3/cigs/cigs.htm.
48. Klepeis NE, Nelson WC, Ott WR, Robinson JP, Tsang AM, Switzer P, Behar JV, Hern SC, Engelmann WH. The National Human Activity Pattern Survey (NHAPS): a resource for assessing exposure to environmental pollutants. Journal of Exposure Analysis and Environmental Epidemiology 2001; 11:231–252.
49. Reponen T, Willeke K, Ulevicius V, Reponen A, Grinshpun SA. Effect of relative humidity on the aerodynamic diameter and respiratory deposition of fungal spores. Atmospheric Environment 1996; 30:3967–3974.
50. ANSI/AHAM AC-1-2006. Method for measuring performance of portable household electric room air cleaners. Washington, DC: Association of Home Appliance Manufacturers; 2006. Available at: http://webstore.ansi.org/RecordDetail.aspx?sku=ANSI%2FAHAM+AC-1-2006.
51. Helbling A, Brander KA, Horner WE, Lehrer SB. Allergy to Basidiomycetes. In: Breitenbach M, Crameri R, Lehrer SB, editor. Fungal allergy and pathogenicity. Basel: S. Karger AG; 2002.
52. Dick EC, Jennings LC, Mink KA, Wartgow CD, Inhorn SL. Aerosol transmission of rhinovirus colds. Journal of Infectious Diseases 1987; 156:442–448.
53. Wells W. Airborne contagion and air hygiene. Cambridge, MA: Harvard University Press; 1955.
54. Rudnick SN, Milton DK. Risk of indoor airborne infection transmission estimated from carbon dioxide concentration. Indoor Air 2003; 13:237–245.
55. Riley RL, Nardell EA. Cleaning the air. The theory and application of ultraviolet air disinfection [Review]. American Review of Respiratory Disease 1989; 139:1286–1294.
56. Liao CM, Chang CF, Liang HM. A probabilistic transmission dynamic model to assess indoor airborne infection risks. Risk Analysis 2005; 25:1097–1107.
57. Loudon RG, Roberts RM. Droplet expulsion from the respiratory tract. The American Review of Respiratory Disease 1967; 95:435–442.
58. Duguid J. The size and duration of air-carriage of respiratory droplets and droplet-nuclei. Journal of Hygiene 1946; 44:471–479.
59. Nicas M, Nazaroff WW, Hubbard A. Toward understanding the risk of secondary airborne infection: emission of respirable pathogens. Journal of Occupational and Environmental Hygiene 2005; 2:143–154.
60. van Elden LJ, Nijhuis M, Schipper P, Schuurman R, van Loon AM. Simultaneous detection of influenza viruses A and B using real-time quantitative PCR. Journal of Clinical Microbiology 2001; 39:196–200.
61. Riley E, Murphy G, Riley R. Airborne spread of measles in a suburban elementary school. American Journal of Epidemiology 1978; 107:421–432.
62. United States Environmental Protection Agency (EPA). Exposure factors handbook. Washington, DC: Office of Research and Development, National Center for Environmental Assessment, U.S. Environmental Protection Agency; 1997. Available at: http://cfpub.epa.gov/ncea/cfm/recordisplay.cfm?deid=12464.
63. Long CM, Suh HH, Catalano PJ, Koutrakis P. Using time- and size-resolved particulate data to quantify indoor penetration and deposition behavior. Environmental Science Technology 2001; 35:2089–2099.
64. Abt E, Suh HH, Catalano P, Koutrakis P. Relative contribution of outdoor and indoor particle sources to indoor concentrations. Environmental Sciency Technology 2000; 34:3579–3587.

65. Spengler J. Long-term measurements of respirable sulfates and particles inside and outside homes. Atmospheric Environment 1981; 15:23–30.
66. Lee T, Grinshpun SA, Martuzevicius D, Adhikari A, Crawford CM, Reponen T. Culturability and concentration of indoor and outdoor airborne fungi in six single-family homes. Atmospheric Environment 2006; 40:2902–2910.
67. Gomes C, Freihaut J, Bahnfleth WP. Resuspension of allergen-containing particles under mechanical and aerodynamic disturbances from human walking. Atmospheric Environment 2007; 41:5257–5270.
68. Myatt TA, Johnston SL, Zuo Z, Wand M, Kebadze T, Rudnick S, Milton DK. Detection of airborne rhinovirus and its relation to outdoor air supply in office environments. American Journal of Respiratory and Critical Care Medicine 2004; 169:1187–1190.
69. Chen SC, Chang CF, Liao CM. Predictive models of control strategies involved in containing indoor airborne infections. Indoor Air 2006; 16:469–481.
70. Fisk W. Commentary on predictive models of control strategies involved in containing indoor airborne infections. Indoor Air 2008; 18:72–73.
71. Ijaz MK, Brunner AH, Sattar SA, Nair RC, Johnson-Lussenburg CM. Survival characteristics of airborne human coronavirus 229E. Journal of General Virology 1985; 66:2743–2748.
72. Reisman RE, Mauriello PM, Davis GB, Georgitis JW, DeMasi JM. A double-blind study of the effectiveness of a high-efficiency particulate air (HEPA) filter in the treatment of patients with perennial allergic rhinitis and asthma. The Journal of Allergy and Clinical Immunology 1990; 85:1050–1057.
73. Nelson HS, Hirsch SR, Ohman JL, Jr, Platts-Mills TA, Reed CE, Solomon WR. Recommendations for the use of residential air-cleaning devices in the treatment of allergic respiratory diseases. Journal of Allergy and Clinical Immunology 1988; 82:661–669.
74. Gore RB, Bishop S, Durrell B, Curbishley L, Woodcock A, Custovic A. Air filtration units in homes with cats: can they reduce personal exposure to cat allergen? Clinical and Experimental Allergy 2003; 33:765–769.
75. National Environmental Education & Training Foundation (NEETF). Environmental management of pediatric asthma. Guidelines for health care providers. Washington, DC: National Environmental Education & Training Foundation (NEETF); 2005.
76. Custovic A, Fletcher A, Pickering CA, Francis HC, Green R, Smith A, Chapman M, Woodcock A. Domestic allergens in public places III: house dust mite, cat, dog and cockroach allergens in British hospitals. Clinical and Experimental Allergy 1998; 28:53–59.

7 Home Humidification and Influenza Virus Survival

Theodore A. Myatt, Matthew H. Kaufman, Joseph G. Allen, David L. MacIntosh, M. Patricia Fabian, and James J. McDevitt

CONTENTS

INTRODUCTION

Laboratory research studies indicate that aerosolized influenza viruses survive for longer periods at low relative humidity (RH) conditions. Further analysis has shown that absolute humidity (AH) may be an improved predictor of virus survival in the environment. Maintaining airborne moisture levels that reduce survival of the virus in the air and on surfaces could be another tool for managing public health risks of influenza.

A multi-zone indoor air quality model was used to evaluate the ability of portable humidifiers to control moisture content of the air and the potential related benefit of decreasing survival of influenza viruses in single-family residences. We modeled indoor AH and influenza virus concentrations during winter months (Northeast US) using the CONTAM multi-zone indoor air quality model. A two-story residential template was used under two different ventilation conditions—forced hot air and radiant heating. Humidity was evaluated on a room-specific and whole house basis. Estimates of emission rates for influenza virus were particle-size specific and derived from published studies and included emissions during both tidal breathing and coughing events. The survival of the influenza virus was determined based on the established relationship between AH and virus survival.

The presence of a portable humidifier with an output of 0.16 kg water per hour in the bedroom resulted in an increase in median sleeping hours AH/RH levels of 11 to 19% compared to periods without a humidifier present. The associated percent decrease in influenza virus survival was 17.5–31.6%. Distribution of water vapor through a residence was estimated to yield 3 to 12% increases in AH/RH and 7.8–13.9% reductions in influenza virus survival.

This modeling analysis demonstrates the potential benefit of portable residential humidifiers in reducing the survival of aerosolized influenza virus by controlling humidity indoors.

BACKGROUND

Annual influenza epidemics exhibit a strong seasonal cycle in temperate regions. Due to the cyclical nature, it has long been assumed that environmental factors played a role in the seasonal epidemics [1]. Numerous laboratory studies have demonstrated that aerosolized influenza virus survival in the air and on surfaces is affected by temperature and more importantly, relative humidity (RH) [2–9]. These studies, carried out using a variety of methods, show that aerosolized influenza virus survives substantially longer at low RH levels. For example, the results of the study conducted by Harper, show that aerosolized influenza survived best when the RH was below 36%, with a sudden decrease in survival of the virus when the RH was raised above 49% [7]. While the data are more limited, studies have also shown RH impacts influenza virus survival on surfaces [10–12].

RH, the ratio of the vapor pressure of water to the saturation vapor pressure at a prescribed temperature and pressure, has been the parameter of interest in past influenza survival and transmission studies. In a reanalysis of influenza survival and transmission data Shaman and Kohn show that compared to RH, absolute humidity (AH), the mass of water per volume of air, has a much stronger statistically significant relationship with influenza virus survival [13]. These results have been extended in an epidemiological model that indicates that AH, as a modulator of influenza transmission, drives seasonal variations of influenza transmission in temperate regions [14].

Indoor moisture levels are dependent on outdoor moisture loads, indoor moisture sources and ventilation rates. Studies conducted in Finland, Canada and Wisconsin have shown that heating season indoor RH levels are low, ranging between 15 and 45% with mean levels of approximately 35% [15–17] which corresponds to 8.1 millibar (mb) AH (at 20°C and standard pressure). The laboratory studies discussed above suggest that increasing the moisture levels above typical indoor AH and RH may mitigate the spread of influenza viruses in the air and on surfaces during the influenza season [13, 18].

The objective of this study was to evaluate the effect of using portable humidifiers on AH levels and influenza virus survival in Northeast United States residences using the CONTAM multi-zone indoor air quality model.

METHODS

The CONTAM multi-zone indoor air quality model (National Institute of Standards and Technology (NIST), Gaithersburg, MD), was used to estimate moisture levels (i.e., RH and AH) and indoor influenza virus concentrations indoors [19]. CONTAM generates dynamic simulations of inter-zonal airflows, ventilation rates, and concentrations of gaseous and aerosol contaminants. The performance of the model has been evaluated extensively. Estimated inter-zonal airflows and air exchange rates have been shown to be within 15% on average of actual measurements and modeled fine particle levels within 30% of measured values [20–23].

Our analysis utilized a two-story 88 m^3 detached residential building template developed by NIST (Template DH-28). Simulations were conducted for two types of heating systems: a forced hot air system and a radiant heating system. The primary difference between the two systems for purposes of this analysis is that forced air systems provide more rapid mixing of indoor air throughout a residence than radiant heating. The forced air system provided 0.18 $m^3/min/m^2$ of air to each room in the house. The air handler duty schedule was simulated with 1-hour resolution based on output from the EnergyPlus Energy Simulation Software [24]. In general, the fraction of each hour devoted to forced air heating was proportional to the difference between ambient temperature and a set point of 17.8°C (64°F) from 10 PM to 5 AM and 22.2°C (72°F) from 5 AM to 10 PM. Additional details of the residential template and heating systems are described elsewhere [25, 26].

Indoor moisture sources and their generation rates were based on published data [27]. Indoor moisture sources included cooking, dishwashing, bathing, and both waking and sleeping occupants. After inclusion of these sources, the model was calibrated to achieve average indoor RH levels reported for homes during the heating seasons [15–17]. A humidifier moisture source with a generation of 0.16 kg/hr (Model V4500, Kaz, Inc., Southborough, MA) was added to either a single bedroom or all the bedrooms and the family room depending on the modeling scenario.

Meteorological information is used by CONTAM to simulate force convection, radiant leakage, and corresponding air exchange rates. We used National Renewable Energy Laboratory TMY2 typical meteorological year meteorological data during the Northern hemisphere influenza season (October to March), including hourly wind direction and speed, dry and wet bulb temperature, relative humidity, and cloud cover data, obtained from the National Weather Service for Boston, Massachusetts (WBAN 14739 Boston Logan International Airport). The Boston area was chosen because, similar to other areas of the Northern US, Boston experiences long periods of cold, dry weather during the winter months and is therefore likely to have low indoor moisture levels.

Our modeling assumed a single influenza case in the bedroom and evaluated the levels of the virus in the bedroom and outside the bedroom. Emission rates for influenza virus were derived from our previous studies with regard to particle counts, particle sizes, and influenza virus RNA concentrations in persons with confirmed flu and were used as input into the CONTAM model [28–30]. In these previous studies, tidal breathing, particle counts and influenza virus RNA concentrations were measured using an Exhalair device (Pulmatrix, Lexington, MA), which records particle counts between 0.3 um and >5 um with an optical particle counter, and collects exhaled breath particles on Teflon filters. In our prior research for coughs, influenza virus RNA was collected with the Gesundheit II device [31], which collects particles by impaction in two size fractions: fine (< 5 μm) and coarse (> 5 μm). Influenza virus RNA was measured in these previous studies by a reverse transcription-quantitative polymerase chain reaction (RT-qPCR) assay. The relationship of 300 viral RNA copies per infective virus particle (determined via cell culture assays), determined in laboratory studies, was used to convert the concentration of influenza virus RNA copies in each particle to infective virus particles per particle size bin [30].

We modeled tidal breathing as a constant emission source of influenza particles per minute in four different particle sizes. Coughing emissions were modeled as one second bursts of infective influenza particles in fine and coarse particle sizes. A frequency of 15 coughing episodes per hour was based on experimental data of subjects with respiratory illness [32]. Influenza emissions were limited to the evening hours (7 PM to 10 AM) as the evening was assumed to be the period in which a person was most likely to be in a bedroom and most likely to be operating a portable humidifier. Emission rate details are presented in Table 1.

To estimate the decrease of influenza virus in the air (i.e., biological decay) due to the increase in AH, the modeled airborne influenza virus levels, which accounts for the physical decay of the viral particles, were adjusted based on laboratory test data originally published by Harper and reanalyzed by Shaman and Kohn which exhibited a strong statistical relationship between AH and one hour loss of live virus [7, 13]. Briefly, the adjustment was based on the regression of log (percent surviving after one hour) and AH (p-value< 0.0001) presented as Figure 3, part F in Shaman and Kohn (2009). Presentations of results with RH assume an indoor temperature of 17.8°C (64°F).

RESULTS

In the models without humidifiers, the median night hour indoor bedroom moisture level was 33% RH (range: 12 to 65% RH) and 35% RH (range: 7 to 75% RH) for the radiant heat and forced air heat models, respectively. The RH levels translate into median AH levels of 7.5 mb for the radiant heat model and 8.2 mb for the forced air heat model.

The addition of a humidifier in one bedroom increased the median night hour bedroom moisture level to 47% RH (median AH: 10.4 mb) for the radiant heat model and 41% (median AH: 9.4 mb) for the forced air heat model. The effect of the single humidifier on RH levels for a typical 24-hour period (November 14 to 15) is depicted in Figureor to activating the humidifier, both models show room RH to be approximately 30%. After the humidifier is turned on (6 pm), RH levels increase to 40–60% RH (approximately 15–20% above baseline) and remain elevated for the duration of humidifier operation. The spike in RH depicted in the figure is due to an outdoor air event, and not related to an increase in moisture generated indoors. In the scenario with humidifiers in the four bedrooms and the family room, the whole house median moisture level increased to approximately 42% (AH: 10.0 mb) for both the radiant and forced air heat models.

TABLE 1 Infective Influenza Virus Emission Rates.

Source type	Particle size (µm)	Infectious influenza viruses per second	Source description
Cough	**2.5**	**0.62**	**15 one second episodes per hour during sleeping hours**
	7.5	0.11	
Tidal breathing	0.4	8.8E-05	Constant emission during sleeping hours
	0.75	1.9E-05	
	2.5	3.3E-06	
	7.5	7.8E-07	

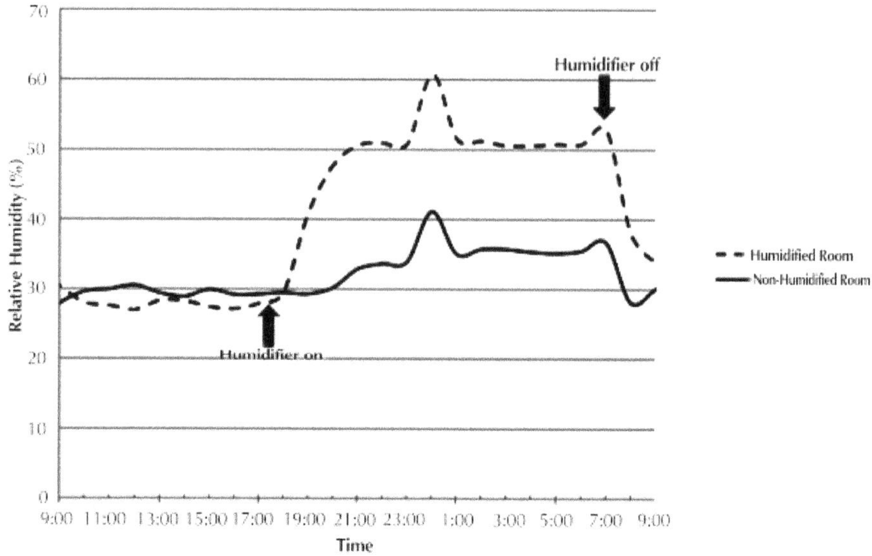

FIGURE 1 Impact of humidifier on typical day (November 14 to November 15) on relative humidity in bedroom with humidifier operating from 6 PM to 7 AM.

Cumulative distributions of modeled hourly bedroom indoor AH levels over the entire modeling period (i.e., October to March) are shown in Figures 2A and 2B. For radiant heat, humidification increased the AH approximately 4 mb, while the increase was approximately one-half as large for the forced air heat.

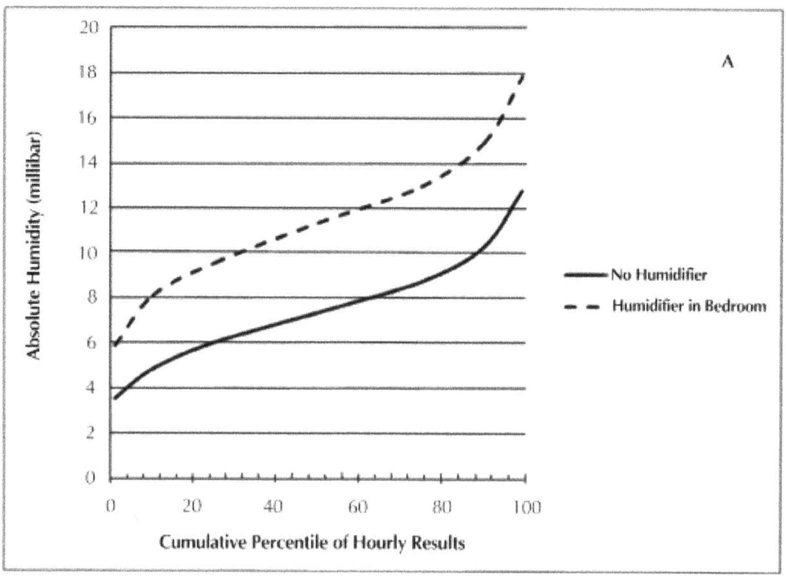

FIGURE 2 *(Continued)*

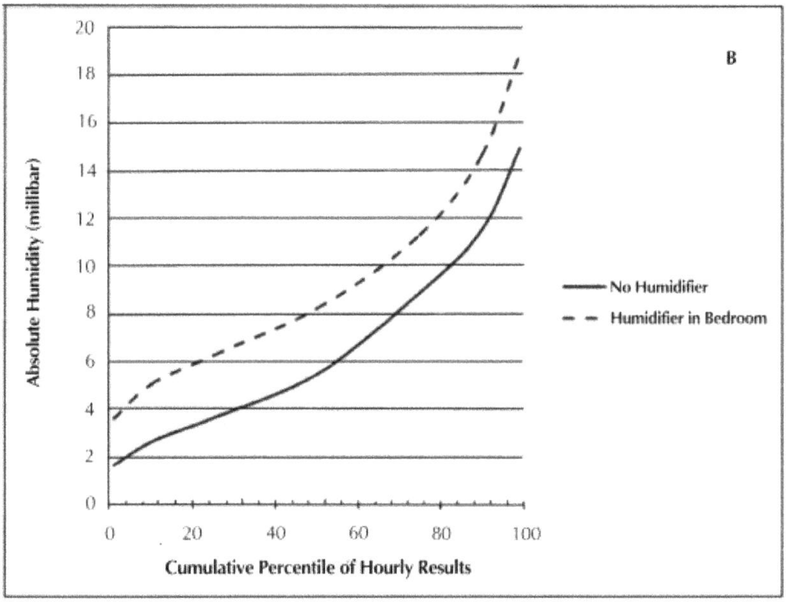

FIGURE 2 Cumulative percentiles of hourly AH comparing rooms with and without humidifiers in the single humidifier in bedroom scenario in (A) model with radiant heat and (B) model with forced-air heat.

The benefit of decreased influenza virus survival due to humidification for the single and multiple humidifier scenarios are presented in Table 2. Estimates of the change in influenza virus survival ranged from 17.5% to 31.6% reduction in rooms with a humidifier operating. The largest decrease in influenza virus survival was in the home with radiant heat due to the larger increase in moisture levels in the room. When multiple humidifiers are employed, the rooms with humidifiers have similar reductions to the single humidifier scenario, but the decrease in virus survival on the entire first and second floors are more modest.

While the reductions in influenza virus survival due to humidification were greater for the radiant heat model, the concentration of viruses surviving in the bedroom was 1.9-fold greater in this model compared to the forced air model. The difference in concentration is due to greater air movement between rooms in the forced air model. Virus concentrations outside the bedroom was substantially lower than in the bedroom, with viral concentrations 27- and 77-fold lower in the hallway compared to the bedroom for the forced air and radiant heat models, respectively.

DISCUSSION

Results from this modeling analysis demonstrate that the use of portable residential humidifiers increases RH and AH to levels that can potentially decrease the survival of airborne influenza virus in a residential setting. This effect is more pronounced in rooms where the humidification is located. While this study evaluated the impacts in

TABLE 2 Decrease in Influenza Virus Survival Due to Humidification, Based on CONTAM Output.

Heating Type	Modeled Domain[1]	No Humidification		Humidification		Decrease in Airborne Virus Survival due to Humidification (% change)
		Median hourly RH[2]	Median hourly AH (VP, mb)[2]	Median hourly RH[2]	Median hourly AH (VP, mb)[2]	
Bedroom Humidifier Scenario						
Forced Air	bedroom	27	5.4	38	7.7	17.5
Radiant		34	6.9	53	10.7	31.6
Multiple Humidifier Scenario[3]						
Forced Air	1st floor	26	5.2	29	5.8	7.8
	2nd floor	27	5.4	33	6.6	8
Radiant	1st floor	33	6.6	40	8.1	9.5
	2nd floor	36	7.3	48	9.7	13.9

[1] 1st floor and 2nd floor are floor level averages
[2] Evening hour median levels (7 PM to 10 AM)
[3] Humidifiers in all bedrooms and family room

a residential setting, the expected benefits of humidification are likely to be larger in places where larger populations of people with the flu and people susceptible to the flu congregate.

Increasing low indoor moisture levels may have benefits beyond reducing survival of the influenza virus. Low RH has been associated with a number of symptoms including dry skin, throat and mucous membranes and eye irritation in office and hospital workers [33–35]. In a home humidification intervention, the authors reported a decrease in dryness of the nose and throat and improved breathing in patients with allergies [36].

While there are apparent benefits to residential humidification for control of influenza virus in the air and on surfaces and temporary relief from cough and cold symptoms, excessive indoor humidity can lead to unattended consequences such as mold and mildew growth [37]. This study shows that meaningful reductions of airborne influenza virus are possible at indoor moisture levels that are generally acceptable in terms of overall indoor air quality. However, the model demonstrated that the RH in the humidified bedroom can exceed 60%, an upper level limit recommended in widely accepted guidance manuals [38], especially in the radiant heat model. In most cases the time above 60% RH were short. However, care should be taken to follow humidifier manufacturer's guidance to ensure that moisture levels are maintained at acceptable levels. The risk of mold and mildew growth can be further reduced by raising the temperatures of surfaces through the addition of insulation or other means on condensing surfaces, such as windows and doors [39].

Our modeling effort focused on airborne influenza virus. One can be exposed to influenza by exposure to contaminated aerosols, large droplets, and direct contact with contaminated secretions or fomites. However, there is disagreement in the scientific community as to the relative importance of the various exposure routes [40–42]. A modeling study by Nicas and Jones, showed that the relative contribution of the airborne route of exposure may be modulated by the viral concentration in the salvia, with the higher the concentration the more likely that the airborne route is an important mechanism of exposure [43]. If the relative contribution of aerosol route of exposure is small, the impacts of humidification at minimizing airborne influenza survival may be similarly small. There is data, however, that indicates that increased humidification decreases survival of influenza virus on surfaces [40–42]. Therefore, while the focus of our modeling effort was on airborne influenza virus, humidification may reduce survival of the virus on household surfaces.

No studies were identified in which RH modeled by CONTAM or other models were used to address the impact of moisture levels on biological contaminants [44–46]. However, previous studies have employed CONTAM to assess humidity levels [45, 46]. Using an earlier version of the CONTAM model, researchers from NIST compared model data with indoor RH measurements in two short time periods. While the data was preliminary, the model and measured results has a reasonable agreement [45], indicating the utility of CONTAM in modeling moisture levels.

This modeling study demonstrated differences in the impacts of humidification between forced air and radiant home heating methods. The median air exchange rate of the models was essentially the same, approximately 0.22 air changes per hours

(ACH). However, due to lower air movement between rooms, radiant heat models showed larger increases in moisture levels, and therefore larger decreases in influenza virus. In a similar way, home characteristics such as the room volume and air exchange rate should be considered when evaluating the type and amount of humidification to be added to a home. ACH observed in our modeling effort are lower than previous modeling efforts primarily due to modeling only the winter months where windows are typically closed. Geographic areas should also be considered when interpreting these results. In areas with lower outdoor AH than the meteorological data we employed from Boston, MA, more humidification may be necessary to have a meaningful impact on indoor AH levels.

CONCLUSIONS

Our results build upon previous efforts to evaluate the impacts of moisture on influenza virus survival [2–9, 13, 47]. These laboratory studies have consistently confirmed that survival of the influenza virus in the air and on surfaces is modulated by moisture levels, with the majority showing the lowest level of survival in the range of 40 to 60% RH. While field studies are necessary to confirm our modeling results, our findings suggest that indoor humidification will increase AH and RH to levels shown to reduce levels of the influenza virus. In this way, humidifiers may be important tools to reduce the survival of influenza virus in the home. The effects of humidification on influenza virus survival, however, should be further evaluated with careful and controlled laboratory and field studies.

KEYWORDS

- **Absolute humidity**
- **Indoor moisture levels**
- **Meteorological information**
- **Multi-zone indoor air quality model**
- **Relative humidity**

REFERENCES

1. Thompson WW, Comanor L, Shay DK. Epidemiology of seasonal influenza: use of surveillance data and statistical models to estimate the burden of disease. Journal of Infectious Disease 2006; 194(Suppl 2):S82–91.
2. Edward D, Elford W, Laidlaw P. Studies of air-borne virus infections. Journal of Hygiene (Lond) 1943; 43:1–15.
3. Lester W Jr. The influence of relative humidity on the infectivity of air-borne influenza A virus. PR8 strain. Journal of Experimental Medicine 1948; 88:361–368.
4. Loosli CG, Lemon HM, Robertson OH, Appel E. Experimental air-borne influenza infection. I. Influence on humidity on survival of virus in air. Proc Exp Biol Med 1943; 531:205–206.
5. Hemmes JH, Winkler KC, Kool SM. Virus survival as a seasonal factor in influenza and polimyelitis. Nature 1960; 188:430–431.

6. Shechmeister IL. Studies on the experimental epidemiology of respiratory infections. III. Certain aspects of the behavior of type A influenza virus as an air-borne cloud. Journal of Infectious Disease 1950; 87:128–132.
7. Harper G. Airborne micro-organisms—Survival test with 4 viruses. Journal of Hygiene (Lond) 1961; 59:479.
8. Hood AM. Infectivity of influenza virus aerosols. Journal of Hygiene (Lond) 1963; 61:331–335.
9. Schaffer FL, Soergel ME, Straube DC. Survival of airborne influenza virus: effects of propagating host, relative humidity, and composition of spray fluids. Archives of Virology 1976; 51:263–273.
10. Edward DG. Resistance of influenza virus to drying and its demonstration on dust. Lancet 1941; 241:664–666.
11. Buckland FE, Tyrrell DA. Loss of infectivity on drying various viruses. Nature 1962; 195:1063–1064.
12. McDevitt J, Rudnick S, First M, Spengler J. The role of absolute humidity on the inactivation of influenza viruses on stainless steel surfaces at elevated temperature. Applied Environmental Microbiology 2010; 76(12):3943–3947.
13. Shaman J, Kohn M. Absolute humidity modulates influenza survival, transmission, and seasonality. Proceedings of the National Academy of Science USA. 2009; 106:3243–3248.
14. Shaman J, Pitzer VE, Viboud C, Grenfell BT, Lipsitch M. Absolute humidity and the seasonal onset of influenza in the continental United States. PLoS biology 2010; 8(3).
15. Kalamees T, Korpi M, Vinha J, Kurnitski J. The effects of ventilation systems and building fabric on the stability of indoor temperature and humidity in Finnish detached houses. Building and Environment 2009; 44:1643–1650.
16. TenWolde A. Ventilation, humidity, and condensation in manufactured houses during winter. ASHRAE Transactions 1994; 100(1) http://www.fpl.fs.fed.us/documnts/pdf1994/tenwo94c.pdf.
17. Weichenthal S, Dufresne A, Infante-Rivard C, Joseph L. Indoor ultrafine particle exposures and home heating systems: a cross-sectional survey of Canadian homes during the winter months. Journal of Expo Sci Environmental Epidemiology 2007; 17:288–297.
18. Lipsitch M, Viboud C. Influenza seasonality: lifting the fog. Proceedings of the National Academy of Science USA 2009; 106:3645–3646.
19. Walton G, Dols WS. CONTAM 2.1 Supplemental user guide and program documentation. Gaithersburg, MD: National Institute of Standards and Technology; 2006.
20. Emmerich S, Nabinger S. Measurement and simulation of the IAQ impact of particle air cleaners in a single-zone building. Gaithersburg, MD: National Institute of Standards and Technology; 2000. http://www.bfrl.nist.gov/IAQanalysis/docs/NISTIR6461.pdf.
21. Emmerich SJ, Nabinger S, Gupte A, Howard-Reed C, Wallace L. Comparison of measured and predicted tracer gas concentrations in a townhouse. Gaithersburg, MD: National Institute of Standards and Technology; 2003. http://www.bfrl.nist.gov/IAQanalysis/docs/NISTIR_7035_final1-CC.pdf.
22. Howard-Reed C, Nabinger S, Emmerich SJ. Predicting the performance of non-industrial gaseous air cleaners: measurements and model simulations from a pilot study. Gaithersburg, MD: National Institute of Standards and Technology; 2004.
23. Lansari A, Streicher J, Huber A, Crescenti G, Zweidinger R, Duncan J, Weisel C, Burton R. Dispersion of automotive alternative fuel vapors within a residence and its attached garage. Indoor Air 1996; 6:118–126.
24. US DOE. EnergyPlus Version 2.1.0. Washington, DC; 2007. http://apps1.eere.energy.gov/buildings/energyplus/features.cfm.
25. Myatt TA, Minigishi T, Allen JG, MacIntosh DL. Control of asthma triggers in indoor air with air cleaners: a modeling analysis. Environmental Health 2008; 7.
26. Persily A, Musser A, Leber D. A collection of homes to represent U.S. housing stocks. Gaithersburg, MD: National Institute of Standards and Technology; 2006.

27. Howard-Reed C, Polidori A. Database tools for modeling emissions and controls of air pollutants from consumer products, cooking and combustion. I. Gaithersburg, MD: National Institute of Standards and Technology; 2006.
28. Fabian P, McDevitt J, DeHaan WH, Brande M, Milton DK. Influenza virus in fine particles exhaled during tidal breathing and coughing. San Diego, CA: American Thoracic Society International Conference, Poster Session; 2009.
29. Fabian P, McDevitt JJ, DeHaan WH, Fung RO, Cowling BJ, Chan KH, Leung GM, Milton DK. Influenza virus in human exhaled breath: an observational study. PLoS ONE 2008; 3:e2691.
30. Fabian P, McDevitt JJ, Lee WM, Houseman EA, Milton DK. An optimized method to detect influenza virus and human rhinovirus from exhaled breath and the airborne environment. Journal of Environmental Monitoring 2009; 11:314–317.
31. Milton DK, Fabian P, Angel M, Perez DR, McDevitt J. Influenza virus aerosols in human exhaled breath: particle size, culturability, and effect of surgical masks. New Orleans, LA: American Thoracic Society. Poster Session; 2010.
32. Loudon RG, Brown LC. Cough frequency in patients with respiratory disease. Am Rev Respir Dis 1967; 96:1137–1143.
33. Reinikainen LM, Jaakkola JJ, Seppanen O. The effect of air humidification on symptoms and perception of indoor air quality in office workers: a six-period cross-over trial. Archives of Environmental Health 1992; 47:8–15.
34. Nordstrom K, Norback D, Akselsson R. Effect of air humidification on the sick building syndrome and perceived indoor air quality in hospitals: a four month longitudinal study. Occup Environ Medicine 1994; 51:683–688.
35. McIntyre DA. Response to atmospheric humidity at comfortable air temperature: a comparison of three experiments. Ann Occup Hyg. 1978; 21:177–190.
36. Sale CS. Humidification during the cold weather to assist perennial allergic rhinitis patients. Ann Allergy 1971; 29:356–357.
37. IOM. Damp indoor spaces and health. Washington, DC: National Academy Press; 2004.
38. ASHRAE. Standard 62.1-2007: Ventilation for acceptable indoor air quality. Atlanta, GA: American Society of Heating, Refrigeration, and Air Conditioning Engineers; 2007.
39. CPSC. Biological pollutants in your home. Bethesda, MD: U.S. Consumer Product Safety Commission; 1997. http://www.cpsc.gov/cpscpub/pubs/425.html.
40. Tellier R. Review of aerosol transmission of influenza A virus. Emerg Infectious Disease 2006; 12:1657–1662.
41. Tellier R. Transmission of influenza A in human beings. The Lancet 2007; 7:759–760, author reply 761–753.
42. Brankston G, Gitterman L, Hirji Z, Lemieux C, Gardam M. Transmission of influenza A in human beings. The Lancet 2007; 7:257–265.
43. Nicas M, Jones RM. Relative contributions of four exposure pathways to influenza infection risk. Risk Analysis 2009; 29:1292–1303.
44. Glass S, TenWolde A. Review of moisture balance models for residential indoor humidity. Proceedings of the 12th Canadian Conference on Building Science and Technology 2009; 1:231–245. http://www.fpl.fs.fed.us/documnts/pdf2009/fpl_2009_glass001.pdf.
45. Emmerich SJ, Persily A, Nabinger S. Modeling moisture in residential buildings with a multizone IAQ program. Gaithersbur, MD: National Institute of Standards and Technology; 2002. http://fire.nist.gov/bfrlpubs/build03/PDF/b03007.pdf.
46. Emmerich SJ, Howard-Reed C, Gupte A. Modeling the IAQ impacts of HHI interventions in inner city housing. Gaithersburg, MD: National Institute of Standards and Technology; 2005. http://www.fire.nist.gov/bfrlpubs/build05/PDF/b05054.pdf.
47. Lowen AC, Mubareka S, Steel J, Palese P. Influenza virus transmission is dependent on relative humidity and temperature. PLoS Pathogens 2007; 3:1470–1476._

8 Airborne Exposure from Common Cleaning Tasks

Anila Bello, Margaret M. Quinn, Melissa J. Perry, and Donald K. Milton

CONTENTS

INTRODUCTION

A growing body of epidemiologic evidence suggests an association between exposure to cleaning products with asthma and other respiratory disorders. Thus far, these studies have conducted only limited quantitative exposure assessments. Exposures from cleaning products are difficult to measure because they are complex mixtures of chemicals with a range of physicochemical properties, thus requiring multiple measurement techniques. We conducted a pilot exposure assessment study to identify methods for assessing short term, task-based airborne exposures and to quantitatively evaluate airborne exposures associated with cleaning tasks simulated under controlled work environment conditions.

Sink, mirror, and toilet bowl cleaning tasks were simulated in a large ventilated bathroom and a small, unventilated bathroom using a general purpose, a glass, and a bathroom cleaner. All tasks were performed for 10 minutes. Airborne total volatile organic compounds (TVOC) generated during the tasks were measured using a direct reading instrument (DRI) with a photo ionization detector. Volatile organic ingredients of the cleaning mixtures were assessed utilizing an integrated sampling and analytic method, EPA TO–17. Ammonia air concentrations were also measured with an electrochemical sensor embedded in the DRI.

Average TVOC concentrations calculated for 10-minute tasks ranged 0.02–6.49 ppm and the highest peak concentrations observed ranged 0.14–11 ppm. TVOC time concentration profiles indicated that exposures above background level remained present for about 20 minutes after cessation of the tasks. Among several targeted VOC compounds from cleaning mixtures, only 2–BE was detectable with the EPA method. The ten minute average 2–BE concentrations ranged 0.30 –21 ppm between tasks. The DRI underestimated 2–BE exposures compared to the results from the integrated method. The highest concentration of ammonia of 2.8 ppm occurred during mirror cleaning.

Our results indicate that airborne exposures from short-term cleaning tasks can remain in the air even after tasks' cessation, suggesting potential exposures to anyone entering the room shortly after cleaning. Additionally, 2–BE concentrations from cleaning could approach occupational exposure limits and warrant further investigation. Measurement methods applied in this study can be useful for workplace assessment of airborne exposures during cleaning, if the limitations identified here are addressed.

BACKGROUND

A growing body of epidemiologic evidence suggests that workers who perform institutional and domestic cleaning are at increased risks for asthma and other respiratory diseases [1–14]. Very few studies to date have carried out quantitative assessment of workplace cleaning exposures [15–18]. Often qualitative exposure data, such as job titles and product types are used to represent exposure in epidemiologic investigations of asthma from cleaning. Quantitative exposure assessments are necessary for investigations of ingredients potentially responsible for respiratory symptoms among cleaning workers and to evaluate exposure-response relationships [19]. A recent review of asthma and cleaning by Zock et al. [20] emphasized the need for quantitative exposure assessment studies.

Airborne exposures from cleaning products are challenging to quantity because they are complex mixtures of ingredients having a range of volatilities and other physicochemical properties and thus require multiple measurement techniques [21, 22]. An additional challenge for exposure studies is to identify methods that can measure short-term and peak exposures, which are important determinants of respiratory symptoms [23, 24].

The type and the frequency of products used depend on the cleaning task. Multiple cleaning tasks may be performed in one room and, for cleaners in institutions like hospitals and schools, the set of cleaning tasks may be performed repeatedly during the day [10, 21]. We therefore designed a task-based assessment that can provide better evaluation of exposure variability, instead of assessing personal exposures using continuous 8–hour time weighted average measurements. Additionally, by using the task as the unit of analysis, one can investigate short term or peak exposures, as determinants of respiratory symptoms. Finally, the results of task-based assessments can assist in the development of questionnaires for estimating cleaning workers' exposures when measurements are not available.

We conducted a task-based exposure assessment study with two main objectives: a) to identify methods for assessing short term, task-based airborne exposures; and b) to evaluate the airborne exposures associated with cleaning tasks simulated under controlled work environment conditions. Results of this work can provide a foundation for developing a quantitative workplace exposure assessment strategy for an epidemiologic investigation.

METHODS

Selection of Cleaning Products

In an earlier study, we identified cleaning products used for common cleaning activities in six hospitals in Massachusetts. Detailed information on products identified and their chemical compositions are described elsewhere [21]. A set of frequently used products was selected for further quantitative exposure characterization, including a glass cleaner, a general–purpose cleaner, and a bathroom cleaner (Table 1). Selection criteria specified that the product must: 1) contain at least one volatile ingredient identified as a potential respiratory hazard based on our previous qualitative assessment [21]; 2) be task specific; 3) be available via commonly used distributers. Material Safety Data Sheets (MSDSs) indicated 2–buthoxyethanol (2–BE) was a major ingredient in all of the products selected, with concentrations ranging from 0.5%–10% by weight in the bulk products (see Table 1). Other volatile ingredients listed on the MSDS for these mixtures were ethanolamine, ethylene glycol, ethanol, and propylene glycol monoethyl ether.

TABLE 1 Ingredients of Cleaning Products Used for Simulation of Cleaning Tasks 1.

Product	Material Safety Data Sheets' (MSDS) ingredients	CAS number	% by weight
Glass cleaner		111-76-2	25-40
1) concentrate	2- Butoxyethanol	107-98-2	5-7
	Propylene glycol monomethyl ether	NA	5-7
	Alcohol ethoxy sulfate	1336-21-6	3-5
	Ammonium hydroxide	64-02-8	1-3
	Tetrasodium ethylenediamine tetraacetate	64-17-5	0.25-1.0

TABLE 1 *(Continued)*

Product	Material Safety Data Sheets' (MSDS) ingredients	CAS number	% by weight
	Ethyl alcohol		
2) ready to use	Ammonium Hydroxide	1336-21-6	3-5
General Purpose cleaner			
1) concentrate	2-Buthoxyethanol	111-76-2	35-45
	Ethanolamine	141-43-5	10-20
	Sodium hydroxide	1310-73-2	1-1.5
2) ready to use	Mono-ethanolamine	141-43-5	1-3
	2-Buthoxyethanol	111-76-2	5-7
Bathroom cleaner			
1) concentrate	2-Buthoxyethanol	111-76-2	25-40
	Secondary alcohol ethoxylate	68131-40-8	10-25
	Ethanolamine	141-43-5	7-10
	Fragrance	NA	3-5
	Tetrasodium ethylenediamine tetraacetate	64-02-8	1-1.5
	N-Alkyl dimethyl benzyl ammonium chloride	68-424-85-1	0.25-1
	Didecyl dimethyl ammonium chloride	7173-51-5	< 0.1
2) ready to use	2-Buthoxyethanol	111-76-2	1-3

Simulations of Cleaning Tasks

Worksite observational and video analyses of cleaning tasks in two hospitals and one university in Massachusetts were conducted. These analyses focused on bathrooms that our previous qualitative assessment recognized as requiring multiple cleaning tasks. We identified workplace practices related to product application methods, worker's physical movements and proximity to cleaning products, average task duration, and typical room dimensions in which the tasks were performed. The findings from these worksite analyses then were used to develop simulations of the cleaning tasks.

Using the products selected, we simulated three types of cleaning tasks: mirror cleaning (with the glass cleaner), sink cleaning (with the general purpose cleaner), and toilet bowl cleaning (with the bathroom cleaner). Products were sprayed and then wiped using paper towels for mirror and sink cleaning; and a brush for toilet bowl cleaning, as commonly done at the worksite.

The main reason for performing simulations was to control task frequency, duration, and environmental conditions such as ventilation and possible interferences from other sources of volatile compounds. Pilot simulations were initially performed to determine the duration of cleaning tasks needed to collect a sufficient amount of analyte to reach the limit of detection (LOD) of the analytical method, while aiming to conduct the tasks within their actual workplace durations. Workplace observations showed that durations between the tasks varied from 3–10 minutes, depending on the surface dirtiness and the number of toilet bowls, sink, or mirrors in one bathroom. After several simulations and measurements with methods described below, the final task duration was determined as 10 minutes. Integrated air sampling was conducted for each task for the entire simulation period. Direct reading measurements were performed at the same time, but also continued after the tasks stopped, in order to evaluate the after-task exposure profiles.

To investigate the feasibility of capturing a wide range of airborne concentrations (representing lower and higher exposures), cleaning tasks were simulated with varying conditions: in a small and large bathroom; with or without ventilation; and with products at different dilution concentrations. It was hypothesized that factors such as the volume of the room, ventilation conditions, concentrations of the volatile ingredient in the products, and amount of the product used per task, would be important exposure determinants. The small bathroom had dimensions typical of a single patient hospital bathroom and the large bathroom had dimensions typical of a public bathroom with three toilet stalls, four sinks and mirrors (Figure 1). The large bathroom was continuously ventilated with an air exchange rate of 5.5 air changes/hour. The small bathroom's ventilation was controlled using the exhaust fan, which was turned off during the simulations. Amount of the product consumed during each task was recorded by weighing the product bottle before and after each task. The doors and windows were kept closed during the simulations and were opened only after cleaning tasks and measurement had stopped. Paper towels used were removed from the bathrooms after cleaning.

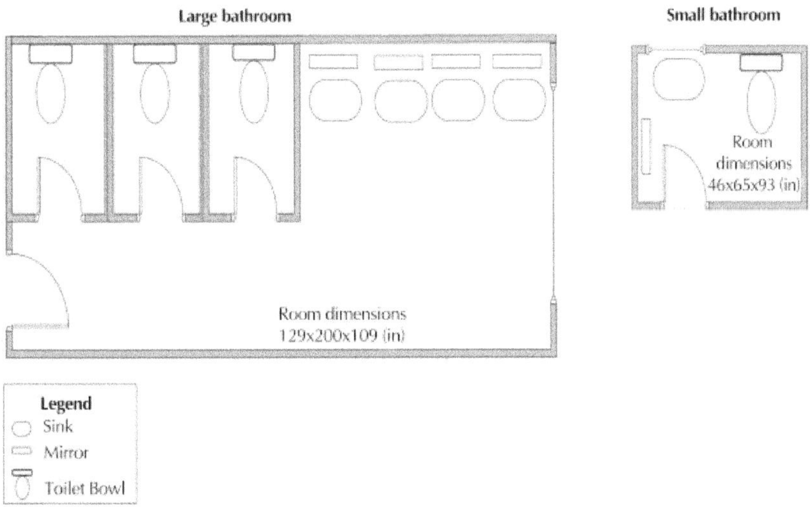

FIGURE 1 Schematic presentation of the bathrooms where cleaning tasks were performed.

Airborne Measurement Methods

Volatile organic compounds were assessed using the following metrics: 1) total volatile organic compounds (TVOC) with direct reading measurements methods; 2) volatile organic compounds (VOC) with a standard integrated sampling and analytical method. Ammonia, a specific ingredient with known respiratory effects, was also assessed concurrently with other VOC metrics. The following measurement methods were used during the simulations:

Direct Reading Measurements of TVOC

Concentrations of TVOC in air were measured using a direct reading instrument (DRI) with a photo ionization detector (PID), Gray Wolf Sensing Solution, the Direct Sense TVOC–TG–502, Trumbull, CT. The PID was equipped with a parts per billion (ppb) sensor with a measurement range of 0.02 –20 parts per million (ppm). Calibration of the ppb sensor was performed at two calibration points: 0 ppm using free air and 7.5 ppm using isobutylene. Concentrations of TVOC were recorded every 15 seconds using a pocket personal computer (PC) connected to the air sampling probe. The data were processed with the Active Sync Software 4.2 and Gray Wolf software version 2.12. The instrument was held constantly in the breathing zone of the person who performed the tasks. Background concentrations of TVOC in the bathrooms were measured before each task. TVOC concentration profiles were obtained during the 10 minutes of the cleaning tasks and continued after their cessation, until the TVOC concentrations dropped to the background level.

Integrated Sampling and Analytical method, EPA TO–17

Integrated sampling was conducted simultaneously with the direct reading TVOC measurements. Breathing zone samples were collected in duplicates on the person who performed the tasks. Active sampling was conducted using the Perkins–Elmer ATD 400 thermal desorption tubes at a flow rate of 65–70 ml/min. Samples were collected continuously for the 10 minutes of the tasks. Following sampling, the tubes were refrigerated and later transported in ice bags for chemical analysis. Compounds sampled were recovered with thermal desorption and analyzed with an Agilent 6890/5973 GC/MS with analytical column J&W DB–1, using helium as the carrier gas.

Ammonia Measures

Ammonia was measured with an electrochemical sensor, which was embedded in the DRI. Similar to TVOC, ammonia concentration-time profiles were obtained during and after each task. The data were recorded and downloaded simultaneously with TVOC using the same software.

RESULTS

TVOC Concentrations

Real time concentration profiles of TVOC for sink, mirror, and toilet bowl cleaning tasks (Figure 2) show TVOC concentrations steadily increasing with time during task performance, reaching the peak at the end of the cleaning period. TVOC concentrations after the tasks declined exponentially to background concentrations. The time to reach the background level typically was about 20 minutes after the tasks had stopped.

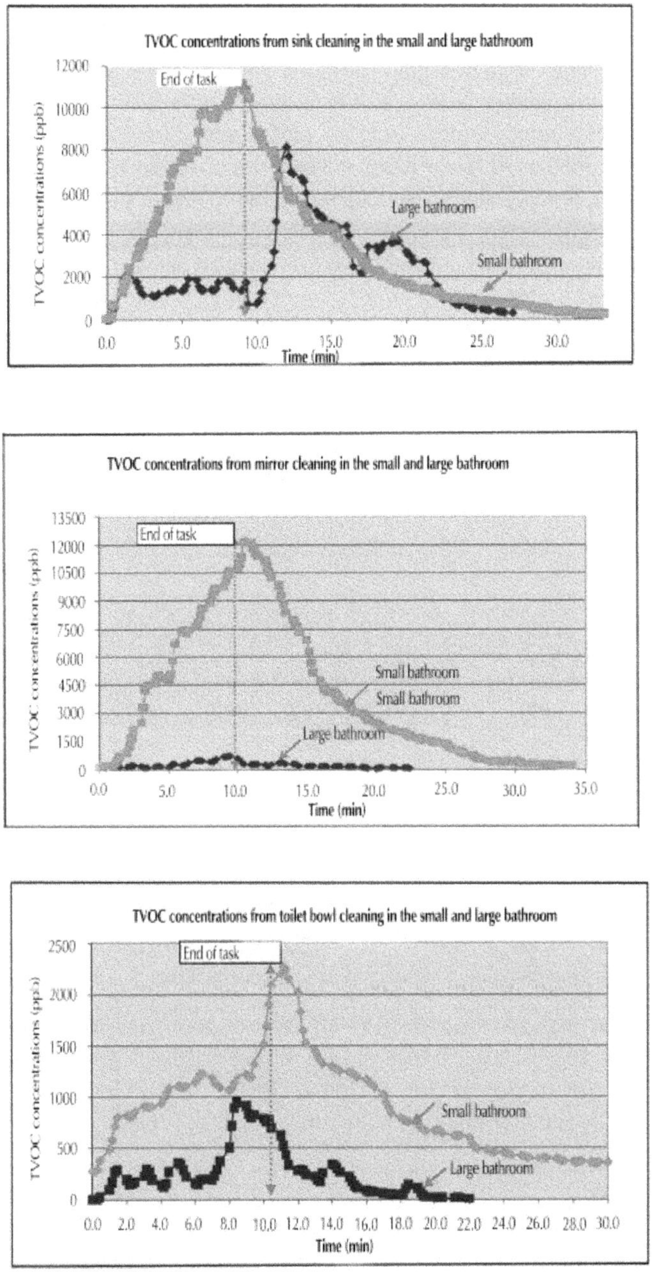

FIGURE 2 Profile of Total Volatile Organic Compounds (TVOC) concentrations during and after cleaning tasks (task duration 10 minutes).

Average TVOC concentrations calculated for 10–minute tasks ranged from 0.02–6.49 ppm and the highest peak concentrations observed for each task ranged from 0.14–11 ppm (Table 2). Overall, concentrations varied by task type, room size, ventilation status, and dilution rate of the product used. Amount of products used did not change much between tasks. The highest peak concentrations were detected during sink and mirror cleaning in the small bathroom without ventilation. Average and peak concentrations were higher in the small bathroom than in the larger one. TVOC profiles show a steady concentration increase in the small bathroom, while in the large bathroom the values were lower and tended to fluctuate more. Variability of TVOC concentrations in the large bathroom can be related to the air mixing from ventilation and movement of the person performing the task from one sink to another. As expected, we found that airborne concentrations were higher when the more concentrated products were used.

Concentrations of Ammonia

The highest peak concentration of ammonia (2.8 ppm) was detected during mirror cleaning (Figure 3), when using the concentrated product that contained 3–5% by weight ammonium hydroxide. Concentration time profiles indicated that ammonia was present even after the tasks had stopped. Lower concentrations were recorded during toilet bowl cleaning (peak of 0.2 ppm), from the product that contained quaternary ammonium compounds at < 1.5% in the concentrated form.

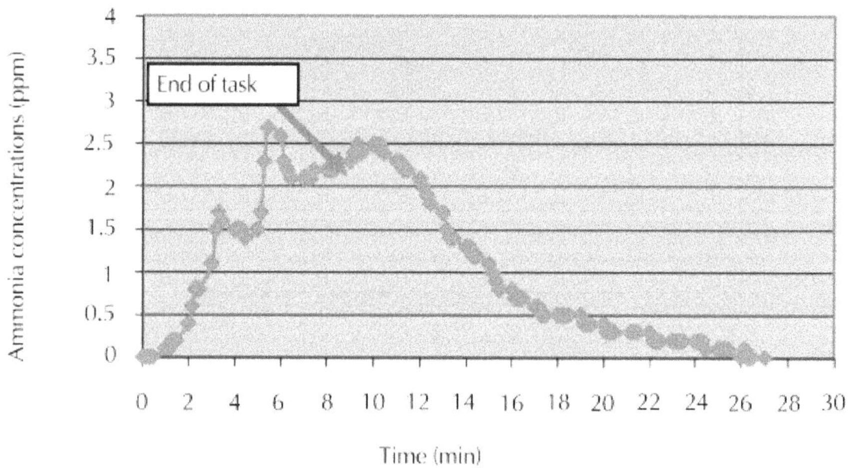

FIGURE 3 Ammonia concentrations profile during and after mirror cleaning in the small bathroom.

Concentrations of 2–Buthoxyethanol

2–BE was the only VOC measured by the EPA TO–17 for the 10 minute sampling. Other target VOC compounds were not detectable from the samples collected. Average concentrations of 2–BE generated from different tasks ranged from 0.3–21 ppm.

Environment/ Task	Product type	Product dilution status	Conc. of 2-BE in the product (% by weight)	Average 2-BE air concentrations in ppm (sd)B	TVOC average concentrations ppm (sd)C	TVOC peak concentrations ppm D
Small Bathroom A						
Sink cleaning	General purpose	Ready to Use (RTU)	5-7	21.27 (2.96)	6.49 (3.56)	11.11
Sink cleaning	General purpose	1 part RTU: 1 water	2.5-3.5	13.32 (2.54)	2.54 (1.51)	4.31
Mirror cleaning	Glass cleaner	1 part concentrated form: 4 parts of water	6-10	13.08 (1.45)	5.26 (3.54)	11.36
Mirror cleaning	Glass cleaner	1 part conc .form: 19 parts of water	1-2	2.96 (0.23)	0.74 (0.39)	1.46
Toilet bowl cleaning	Bathroom cleaner	Ready to Use (RTU)	1-3	3.74 (0.36)	0.96 (0.28)	2.2
Toilet bowl cleaning	Bathroom cleaner	1 part RTU: 1 water	0.5-1.5	2.70 (0.34)	0.56 (0.16)	0.71

TABLE 2 C

Environment/ Task	Product type	Product dilution status	Conc. of 2-BE in the product (% by weight)	Average 2-BE air concentrations in ppm (sd)B	TVOC average concentrations ppm (sd)C	TVOC peak concentrations ppm D
Small Bathroom A						
Large Bathroom						
Sink cleaning	General purpose	Ready to Use (RTU)	5-7	6.27 (0)	1.36 (0.51)	2.13
Sink cleaning	General purpose	1 part RTU: 1 water	2.5-3.5	3.30 (0.1)	0.61 (0.30)	1.37
Mirror cleaning	Glass cleaner	1 part concentrated form: 4 parts of water	6-10	1.98 (0.12)	0.30 (0.19)	0.74
Mirror cleaning	Glass cleaner	1 part conc .form: 19 parts of water	1-2	0.32 (0.01)	0.02 (0.03)	0.14
Toilet bowl cleaning	Bathroom cleaner	Ready to Use (RTU)	1-3	3.05 (0.81)	0.32 (0.27)	0.95
Toilet bowl cleaning	Bathroom cleaner	1 part RTU: 1 water	0.5-1.5	2.76 (0.45)	0.11 (0.08)	0.29

A) Tasks were performed without using the bathroom fan

B) Each value is the average of two side-by side breathing zone samples collected and analyzed according integrated sampling method EPA TO-17

C) The 10 minute average TVOC concentrations measured using a photo-ionization detector (PID)

D) The highest TVOC concentrations recorded by the PID during 10 min tasks.

Airborne concentrations were higher: when products with higher percentage of 2–BE were used; during sink and mirror cleaning compared to toilet bowl cleaning; and when tasks were performed in the small bathroom compared to the large one. The highest concentrations were measured during sink cleaning, when the general purpose cleaner containing 5–7% 2–BE was used.

Correlation of TVOC with 2–BE

Contrary to expectations, the TVOC measurements were consistently lower than 2–BE, a single volatile organic compound of the mixture (Table 2). However, good correlation was found between TVOC and 2–BE values measured simultaneously for the same task ($R^2 = 0.94$).

DISCUSSION

In this study we assessed quantitatively airborne exposures generated from cleaning tasks performed under controlled work environment conditions. Several exposure measures such as TVOC, 2–butoxyethanol (2–BE) and ammonia were assessed with selected measurement methods. Our results show that VOC exposures remain airborne even after the cessation of cleaning tasks, suggesting potential exposure to anyone entering the room shortly after cleaning. Additionally, the results indicated that 2–BE peak concentrations from cleaning can approach occupational exposure limits, warranting further workplace investigations. The quantitative exposure measurements reported here contribute to the limited workplace exposure data in the literature related to cleaning. The main conclusions are discussed below:

1. The measurement methods we applied for assessing exposures from cleaning tasks are useful for future studies, with limitations.

We utilized several measurement methods for quantifying different exposure metrics including integrated sampling and analytical method of EPA TO–17. This method was selected because it provides detection of VOC at low concentrations and grants collection of compounds with a wide range of volatilities (e.g. ethanol BP = 78°C; 2–BE BP = 218°C) by utilizing multi–media sorbent sampling tubes. However, for the 10 minute sampling only 2–BE was detectable with the method. The initial list of target volatile ingredients from the MSDS data included ethanol, 1–methoxy 2–propanol, and ethanolamine. Ethanolamine was removed from the target list because it was not amenable to the method. Ethanol &1–methoxy 2–propanol concentrations were lower than the LOD, for the 10–minute sampling period. For longer sampling periods (such as 20 minutes, data not shown) ethanol, 1–methoxy 2–propane were detectable with the EPA method. Given that our goal of capturing a range of VOC with this method was not achieved (either because of short term sampling or low product concentrations of target compounds), specific measurement of individual compounds may be more feasible to apply for workplace exposure assessment. For example, 2–BE can be measured using the NIOSH 1430 method.

TVOC measured with the DRI–PID underestimated exposures from cleaning activities. Given that the TVOC metric represents the sum of the volatile compounds of the mixture, including 2–BE, one would expect that the value of TVOC would be higher than the single ingredient. Because the 2–BE ionization potential (IP) is lower than the IP of the PID lamp (IP for 2–BE is 8.6 eV vs. 10.6 eV), we expected that 2–BE would be measured by the PID. However, our data indicate a clear 2–BE exposure underestimation by the PID. A possible explanation may be related to the differences in sampling methods between integrated sampling, which is based on active sampling; and the real time sampling, which is based on diffusion. Aerosol particles generated during product spraying may be captured by active sampling and not by the PID, therefore producing higher 2–BE levels by integrated sampling.

The same underestimation of VOC from the PID was observed by Coy et al. [25]. This study compared PID results with integrated sampling during simultaneous measurements from the same solvent mixtures. The authors suggest that PID response underestimation is related to: a) different ionization potential of the individual compounds of the mixture; b) non-linearity of the PID response for high concentrations (2000 ppm); c) the size of the ionization chamber. Consistent with our findings, this study showed high correlation of the PID response with integrated sampling measures.

Due to the observed underestimation, we recommend that TVOC–DRI measurement for cleaning mixtures be conducted only when the limitations are taken into account. DRI can be used for: a) initial screening of TVOC concentrations; b) for evaluating exposure control strategies; and c) for identifying exposure peaks and exposure dynamics, which can be useful for prioritizing activities for further and more precise quantitative measures.

2. Quantification of airborne exposures from cleaning requires investigation of other exposure metrics and a variety of sampling and analytical techniques.

In addition to the exposure metrics considered here (TVOC, 2–BE, ammonia) other metrics can be considered for a comprehensive quantitative exposure assessment strategy. These include assessments of additional chemical agents with important health relevance for respiratory irritation and sensitization, such as quaternary ammonium compounds (quats), ethanolamines, and phenols. Quantitative assessment of these ingredients cannot be achieved with one single method, given that these chemicals have different chemical and physical properties. For example quats can be measured with Ion Chromatography [26] and ethanolamines can be measured with GC/FID NIOSH 2007 method.

Further considerations for quantitative assessment can include aerosol exposure characterizations. Product spraying, a common activity during cleaning, generates liquid aerosols of variable chemical composition, including non-volatile compounds such as quats that have been associated with asthma symptoms in several case reports [27]. However, current literature lacks the evidence on size distributions of particles generated from spraying during cleaning. Determination of respirable or ultrafine particle concentrations from spraying may provide a better understanding of cleaning related health effects. Additionally, there are no studies to date that focus on assessment of aerosol dust particles present in indoor environments as potential carriers of volatile and nonvolatile ingredients from cleaning. Secondary emissions generated from cleaning chemicals reaction with ozone, which have been investigated by experimental studies [17], may also be important to consider when developing quantitative workplace exposure assessment strategies.

3. The quantitative findings for airborne TVOC and 2–BE suggest that common cleaning tasks contribute to poor indoor quality and may present a risk of adverse health effects.

The highest TVOC peak concentrations (approximately 11 ppm) and 10-minute average concentrations (approximately 6 ppm) were measured when the general-purpose cleaner was used in the small unventilated bathroom. Although occupational and environmental standards for indoor air TVOC have not been established, Molhave et al. [28, 29] proposed indoor TVOC concentrations of increasing concern for health effects as follows: a comfort range (< 0.2 mg/m^3), a multi factorial exposure range (0.2–3 mg/m^3); a discomfort range (3–25 mg/m^3); and a toxic range (> 25 mg/m^3). Our peak TVOC concentration data converted to mg/m^3 (isobutylene equivalent) ranged from 0.66–26 mg/m^3. Concentrations we recorded for most of the tasks fall into the discomfort range. These results suggest that cleaning can make a significant contribution to the poor indoor air quality. Additionally, Molhave et al. [29] recommended that if a direct reading detector indicates concentrations above 0.3 mg/m^3, further detailed exposure assessments for health effects evaluations are essential.

TVOC concentrations have been measured in several indoor environments including offices, schools, homes, and hospitals [29–34]. In these settings, the TVOC ranged from 1–25 mg/m^3 and were expressed as the average values for different time durations, from hours to days of air sampling. Even though these studies recognize the possibility of higher short-term TVOC exposures, peak TVOC-activity specific data, which are important for asthma assessment, have not been evaluated [23]. Because the degree of cleaning contribution to the short-term peak exposures is unknown, further assessments in the workplace are needed.

Airborne concentrations from cleaning 2–BE may be a concern in the workplace. Concentrations of 2–BE measured here ranged widely among the tasks, with the highest values obtained when the general purpose cleaner with 5–7% 2–BE by weight was used in the small bathroom, approximately 21 ppm. California Proposition 65 has set the Reference Exposure Limit (REL) for 2–BE at 2.9 ppm for one hour of exposure. Our 2–BE results suggest that application of a general purpose cleaner continuously for several consecutive tasks in the workplace can easily result in worker's exposure higher than the California REL limit.

Several laboratory emissions studies have measured 2–BE concentrations from cleaning products. Slightly lower air concentrations than ours were reported by Zhu et al. [35] in an experimental study that determined 2–BE emission factors using a field and laboratory emission cell (FLEC). One–hour concentrations of 2–BE ranged from 2.8–62 mg/m^3 (0.57–12.6 ppm). Singer et al. [18] investigated emission profiles of 2–BE from several cleaning products and reported concentrations of 0.33–2.3 mg/m^3 over one hour of exposure.

There are very limited workplace exposure data of 2–BE from cleaning. Occupational standards for 2–BE such as OSHA Permissible Exposure Limit (PEL) of 8 hr TWA is 50 ppm and NIOSH Recommended Exposure Limit (REL) for 10 hr TWA is 5 ppm (24 mg/m^3). Vincent and coworkers in 1993 [16] assessed 2–BE workplace exposures for 29 cleaning workers, which ranged from 0.1–7.33 ppm for 8 hour TWA. 2–BE exposures from cleaning may meet OSHA regulations, however, compliance does not always imply that workers are protected from respiratory irritation symptoms from short–term peak exposures [23, 36]. Our findings of concentrations as high as 21 ppm, although not directly comparable with the occupational standards, warrant further assessment of 2–BE from cleaning in the workplace.

Concentrations from the tasks performed (0.01–2.8 ppm) were low compared to OSHA –PEL 8 hour TWA of 50 ppm and NIOSH short-term exposure limit (STEL) 15–min TWA of 35 ppm. Several studies have associated inorganic gases such as ammonia and chlorine with irritation symptoms reported among cleaning workers [17, 15]. Concentrations of ammonia reported by Ramon and coworkers range from 0.6–6.4 ppm with peaks over 50 ppm during domestic cleaning tasks [15]. Lower concentrations were reported by Fedoruk et al. [37] when assessing airborne ammonia from a window and a bathroom tile cleaner. This study concluded that standard cleaning solutions are unlikely to produce significant ammonia exposures, but the authors advise that application of more concentrated products (e.g. > 3%) in poorly ventilated areas may be of concern.

4. Concentrations of TVOC measured after cleaning suggests that exposures may affect not only workers involved in cleaning but also other building occupants.

Real time TVOC concentration profiles after the cessation of cleaning tasks indicated that it takes more than 20 minutes after cleaning for exposures to decline to background levels. This finding relates to a single application of one product used during one task, especially in the small, unventilated room. It would be expected that multiple tasks performed consecutively would generate higher exposure concentrations requiring longer decay times. These results suggest that not only workers involved with cleaning, but others, who are present in the room after cleaning, are potentially exposed. Several emissions studies conducted in laboratory chambers have suggested that ingredients in cleaning products such as glycol ethers are slowly released in the air even hours after product applications [17, 18]. These experimental results indicate that there is a potential risk for exposure to other building occupants not involved with cleaning. Further quantitative investigation in real world scenarios is critical to evaluate airborne exposures after cleaning.

CONCLUSIONS

Measurement methods reported here can be used for workplace assessments of airborne exposures generated during cleaning tasks, if the limitations are addressed. Combinations of individual measurements methods for ingredients of significant health relevance with TVOC direct reading measurements can provide complimentary evidence for an epidemiologic investigation and for developing workplace controls. Additional exposure metrics quantified using a variety of sampling and analytic methods will be needed for more comprehensive quantitative exposure assessment.

Our work also shows that airborne VOC exposures occur during short-term cleaning tasks and that these can remain in the air after the task stops, suggesting potential exposure to anyone entering the room shortly after cleaning. In addition, 2–BE peak concentrations from cleaning could approach occupational exposure limits and warrant further investigation. We recognize that cleaning tasks performed at actual worksites are likely to differ from our simulated tasks in several ways: 1) the duration of tasks is more variable; 2) tasks are performed consecutively in one room (e.g. mirror, sink, and toilet all in one bathroom); and 3) the cleaning task cycle is repeated multiple times in institutions such as hospitals and schools where numerous bathrooms are cleaned in a single day. Due to these differences, workplace exposure concentrations are likely to be different than the values reported here, however these data and the methods used to obtain them can be used as groundwork for conducting a comprehensive quantitative exposure assessment for an epidemiologic investigation.

KEYWORDS

- **Direct reading instrument**
- **Limit of detection**
- **Material safety data sheets**
- **Total volatile organic compounds**
- **Volatile organic compounds**

REFERENCES

1. Kogevinas M, Anto JM, Soriano JB, Tobias A, Burney P. The risk of asthma attributable to occupational exposure. A population-based study in Spain. American J Respiratory Crit Care Med 1996; 154:137–143.
2. Kogevinas M, Anto JM, Sunyer J, Tobias A, Kromhout H, Burney P. Occupational asthma in Europe and other industrialized areas: a population-based study. Lancet 1999; 354:166.
3. Zock J, Kogevinas M, Sunyer J, Almar E, Muniozguren N, Payo F, Sanchez J, Anto JM. Asthma risk, cleaning activities and use of specific cleaning products among Spanish indoor cleaners. Scandinavian Journal of Work, Environment and Health 2001; 27:76–81.
4. Zock J, Kogevinas M, Sunyer J, Jarvis D, Toren K, Anto J. Asthma characteristics in cleaning workers, workers in other risk jobs and office workers. European Respiratory Journal 2002; 20:679–685.
5. Medina–Ramon M, Zock JP, Kogevinas M, Sunyer J, Anto JM. Asthma symptoms in women employed in domestic cleaning: a community based study. Thorax 2003; 58:950–954.
6. Jaakkola JJ, Jaakkola MS. Professional cleaning and asthma. Curr Opin Allergy Clin Immunol 2006; 6:85–90.
7. Delclos GL, Gimeno D, Arif AA, Burau KD, Carson A, Lusk C, Stock T, Symanski E, Whitehead LW, Zock JP, et al. Occupational risk factors and asthma among health care professionals. American Journal of Respiratory Critical Care Medicine 2007; 175:667–675.
8. Zock JP, Plana E, Jarvis D, Anto JM, Kromhout H, Kennedy SM, Kunzli N, Villani S, Olivieri M, Toren K, et al. The use of household cleaning sprays and adult asthma: an international longitudinal study. American Journal of Respiratory Critical Care Medicine 2007; 176:735–741.
9. Kogevinas M, Zock JP, Jarvis D, Kromhout H, Lillienberg L, Plana E, Radon K, Toren K, Alliksoo A, Benke G, et al. Exposure to substances in the workplace and new–onset asthma: an international prospective population–based study (ECRHS–II). Lancet 2007; 370:336–341.
10. Arif AA, Hughes PC, Delclos GL. Occupational exposures among domestic and industrial professional cleaners. Occupational Medicine (Lond) 2008; 58:458–463.
11. Delclos GL, Gimeno D, Arif AA, Benavides FG, Zock JP. Occupational exposures and asthma in health–care workers: comparison of self-reports with a workplace-specific job exposure matrix. American Journal of Epidemiology 2009; 169:581–587.
12. Lynde CB, Obadia M, Liss GM, Ribeiro M, Holness DL, Tarlo SM. Cutaneous and respiratory symptoms among professional cleaners. Occupational Medicine (Lond) 2009; 59:249–254.

13. Obadia M, Liss GM, Lou W, Purdham J, Tarlo SM. Relationships between asthma and work exposures among non–domestic cleaners in Ontario. American Journal of Ind Medicine 2009; 52:716–723.
14. Zock JP, Plana E, Anto JM, Benke G, Blanc PD, Carosso A, Dahlman–Hoglund A, Heinrich J, Jarvis D, Kromhout H, et al. Domestic use of hypochlorite bleach, atopic sensitization, and respiratory symptoms in adults. Journal of Allergy Clinical Immunology 2009; 124:731–738.
15. Medina–Ramon M, Zock JP, Kogevinas M, Sunyer J, Torralba Y, Borrell A, Burgos F, Anto JM. Asthma, chronic bronchitis, and exposure to irritant agents in occupational domestic cleaning: a nested case–control study. Occupational and Environmental Medicine 2005; 62:598–606.
16. Vincent R, Cicolella A, Surba I, Reieger B, Poirot P, Pierre F. Occupational exposure to 2–butoxyethanol for workers using window cleaning agents. Applied Occupational and Environmental Hygiene 1993; 8:580–586.
17. Nazaroff WW, J WC. Cleaning products and air fresheners: exposure to primary and secondary air pollutants. Atmospheric Environment 2004; 38:2841–2865.
18. Singer BC, Destaillats H, Hodgson AT, Nazaroff WW. Cleaning products and air fresheners: emissions and resulting concentrations of glycol ethers and terpenoids. Indoor Air 2006, 16:179–191.
19. Rappaport SM, Kupper LL. Quantitative exposure assessment. El Cerrito, CA: Stephen Rappaport, 2008.
20. Zock JP, Vizcaya D, Le Moual N. Update on asthma and cleaners. Curr Opin Allergy Clin Immunol 2010; 20.
21. Bello A, Quinn MM, Perry MJ, Milton DK. Characterization of occupational exposures to cleaning products used for common cleaning tasks—a pilot study of hospital cleaners. Environ Health 2009; 8:11.
22 Wolkoff P, Schneider T, Kildeso J, Degerth R, Jaroszewski M, Schunk H. Risk in cleaning: chemical and physical exposure. Science of Total Environment 1998; 215:135–156.
23 Brooks SM, Hammad Y, Richards I, Giovinco–Barbas J, Jenkins K. The spectrum of irritant–induced asthma: sudden and not–so–sudden onset and the role of allergy. Chest 1998; 113:42–49.
24 Medina–Ramon M, Zock JP, Kogevinas M, Sunyer J, Basagana X, Schwartz J, Burge PS, Moore V, Anto JM. Short-term respiratory effects of cleaning exposures in female domestic cleaners. European Respiratory Journal 2006; 27:1196–1203.
25. Coy JD, Bigelow PL, Buchan RM, Tessari JD, Parnell JO. Field evaluation of a portable photoionization detector for assessing exposure to solvent mixtures. American Industrial Hygiene Association Journal 2000; 61:268–274.
26. Vincent G, Kopferschmitt–Kubler MC, Mirabel P, Pauli G, Millet M. Sampling and Analysis of Quaternary Ammonium Compounds (QACs) Traces in Indoor Atmosphere. Environmental Monitoring and Assessment 2006; 16:16.
27. Purohit A, Kopferschmitt–Kubler MC, Moreau C, Popin E, Blaumeiser M, Pauli G. Quaternary ammonium compounds and occupational asthma. Int Arch Occup Environ Health 2000; 73:423–427.
28. Molhave L. Volatile organic compounds, indoor air quality and health. Indoor Air 1991; 4:357-376.
29. Molhave L, Clausen G, Berglund B, DeCeaurriz J, Kettrup A, Lindvall T, Maroni M, Pickering A, Risse U, Rothweiler H, et al. Total volatile organic compounds (TVOC) in indoor air quality investigations. Indoor Air 1997; 7:225–240.
30. Andrersson K, Bakke J, Bjorseth O, Bornehag C, Clausen G, Hongslo JK, Kjellman M, Kjaergaard S, Levy F, Molhave L, et al. TVOC and health in non-industrial indoor environments. Indoor Air 1997; 7:78–91.
31. Brown SK, Sim MR, Abramson MJ, Gray CN. Concentrations of volatile organic compounds in indoor air—a review. Indoor Air 1994; 4:123–134.
32. Brinke Ten J, Selvin S, Hodgson TA, Fisk JW, Mendell JM, Koshland PC, Daisey MJ. Development of new volatile organic compound (VOC) exposure metrics and their relationship to "sick building syndrome" symptoms. Indoor Air 1998; 8:140–152.

33. Hodgson AT. A review and a limited comparison of methods for measuring total volatile organic compounds in indoor air. Indoor Air 1995; 5:247–257.
34. Wallace L, Pellizzari E, Wendel C. Total volatile organic concentrations in 2700 personal, indoor, and outdoor air samples collected in the US EPA team studies. Indoor Air 1991; 4:465–477.
35. Zhu J, Cao XL, Beauchamp R. Determination of 2–butoxyethanol emissions from selected consumer products and its application in assessment of inhalation exposure associated with cleaning tasks. Environ Int 2001; 26:589–597.
36. Nielsen GD, Wolkoff P, Alarie Y. Sensory irritation: risk assessment approaches. Regul Toxicol Pharmacol 2007; 48:6–18.
37. Fedoruk MJ, Bronstein R, Kerger BD. Ammonia exposure and hazard assessment for selected household cleaning product uses. J Expo Anal Environ Epidemiol 2005; 15:534–544.

9 Phthalate Monoesters from Personal Care Products

Susan M. Duty, Robin M. Ackerman, Antonia M. Calafat, and Russ Hauser

CONTENTS

INTRODUCTION

Phthalates are multifunctional chemicals used in a variety of applications, including personal care products. The present study explored the relationship between patterns of personal care product use and urinary levels of several phthalate metabolites. Subjects include 406 men who participated in an ongoing semen quality study at the Massachusetts General Hospital Andrology Laboratory between January 2000 and February 2003. A nurse-administered questionnaire was used to determine use of personal care products, including cologne, aftershave, lotions, hair products, and deodorants. Phthalate monoester concentrations were measured in a single spot urine sample by isotope dilution–high-performance liquid chromatography coupled to tandem mass spectrometry. Men who used cologne or aftershave within 48 hr before urine collection had higher median levels of monoethyl phthalate (MEP) (265 and 266 ng/mL, respectively) than those who did not use cologne or aftershave (108 and 133 ng/mL, respectively). For each additional type of product used, MEP increased 33% (95% confidence interval, 14–53%). The use of lotion was associated with lower urinary levels of monobutyl phthalate (MBP) (14.9 ng/mL), monobenzyl phthalate (MBzP) (6.1 ng/mL), and mono(2-ethylhexyl) phthalate (MEHP) (4.4 ng/mL) compared with men who did not use lotion (MBP, 16.8 ng/mL; MBzP, 8.6 ng/mL; MEHP, 7.2 ng/mL). The identification of personal care products as contributors to phthalate body burden is an important step in exposure characterization. Further work in this area is needed to identify other predictors of phthalate exposure.

Phthalates are used industrially as plasticizers and solvents and as stabilizers for colors and fragrances. They are found in personal care products, medications, paints, adhesives, and medical equipment made with polyvinyl chloride plastics [1–3]. Diethyl phthalate (DEP), di(2-ethylhexyl) phthalate (DEHP), butylbenzyl phthalate (BBzP), and di-n-butyl phthalate (DBP) are used in personal care products [4, 5]. The potential effects of phthalates on human health are not well characterized. There is a paucity of existing data describing phthalate-associated human health outcomes, although animal studies have found testicular toxicity associated with phthalate exposure [6, 7].

Two studies provide preliminary evidence of associations between urinary concentrations of monoethyl phthalate (MEP), a metabolite of DEP, and DNA damage in human sperm [8], as well as relationships of monobutyl phthalate (MBP) and monobenzyl phthalate (MBzP) phthalate, metabolites of DBP and BBzP, respectively, with decreased sperm motility [9]. In a recent epidemiologic study prenatal exposure to MEP, MBP, MBzP, and monoisobutyl phthalate was associated with shortened anogenital distance (AGD) in male infants [10]. In rodent studies AGD is a sensitive measure of prenatal antiandrogen exposure.

Despite the recent public and scientific interest on the potential human health effects of phthalates, routes of human exposure to phthalates have not been adequately characterized. Potential routes include dietary ingestion of phthalate-containing foods, inhalation of indoor and outdoor air, and dermal exposure through the use of personal care products that contain phthalates. As far as we know, the proportional contribution of phthalate-containing personal care products to total body burden has not been studied. Houlihan et al. [4] quantified phthalate levels in 72 personal care products obtained at a supermarket in the United States, including hair gel/hair spray, body lotion, fragrances, and deodorant. DEP was detected in 71% of these products, DBP in 8%, BBzP in 6%, and DEHP in 4% of the products tested [4]. In a recent study [5], high-performance liquid chromatography (HPLC) was used to quantify the levels of the same four phthalates in 102 hair sprays, perfumes, deodorants, and nail polishes purchased at retail stores in Seoul, Korea. DBP was detected in 19 of the 21 nail polishes and in 11 of the 42 perfumes; DEP was detected in 24 of the 42 perfumes and 2 of the 8 deodorants.

The assertion that phthalates are absorbed into the circulation through human skin is physiologically plausible and is supported by a limited number of human and animal studies [1–3]. The stratum corneum of the epidermis regulates transdermal absorption, and uptake is achieved through passive diffusion [11]. Water-soluble substances penetrate hydrolyzed keratin, whereas lipid-soluble substances such as phthalates, especially DEP and other low-molecular-weight phthalates, can dissolve into lipid materials between keratin filaments. After penetration of the epidermis, diffusion into the dermal and subcutaneous layers is generally uninhibited because of the nonselective and porous aqueous mediums in these layers. Substances can then enter the systemic circulation through venous and lymphatic capillaries. With increased hydration, rates of absorption of more hydrophilic compounds can be increased 3–5 times more than usual. Epidermal permeability also varies greatly between species [11].

In one study dermal doses of DEHP were administered to human volunteers over a 24-hr period, and approximately 1.8% of the total dose was absorbed [12]. Another experiment involved the topical application of DBP to human volunteers. The authors determined that 68 mg would be absorbed in 1 hr if the skin surface of the whole body were saturated with the chemical [13]. In another study human breast skin was exposed in vitro to 14C-DEP, and average absorption under conditions of occlusion was 3.9% compared with 4.8% without occlusion at 72 hr. However, this was much slower and less complete compared with absorption through rat skin [14].

In vitro and animal experiments have also indicated that phthalates are absorbed percutaneously [14–20]. However, the mechanism explaining differential rates of uptake is not agreed upon. Scott et al. (1987) attributed the phthalate-specific rates of absorption to varying degrees of lipophilicity. Elsisi et al. [17] observed that the lengths of the alkyl chains were inversely associated with the relative rates of absorption; except for dimethyl phthalate, DEP has the shortest alkyl chain [1].

Although diester and monoester phthalates have short biologic half-lives of approximately 6–12 hr and do not accumulate [1–3], the frequent application of personal care products may result in semi-steady-state levels, making it possible to estimate typical phthalate body burden from a single urine sample [21, 22]. After exposure, diester phthalates, which may be found in personal care products, are metabolized to monoester metabolites, the suspected toxic agents [6]. For this reason and to avoid contamination, monoester phthalate metabolites rather than the parent diesters are commonly measured [23].

Our objective in the present study was to determine whether the use of personal care products predicted urinary levels of phthalate monoesters, and to identify subject characteristics that predicted phthalate levels.

MATERIALS AND METHODS

Design and Setting

This study was approved by the Human Subject Committees at the Harvard School of Public Health, Massachusetts General Hospital (MGH), and Simmons College. All subjects signed an informed consent. Subjects were participants in an ongoing study on phthalates and male reproductive health. They were recruited between January 2000 and February 2003 from the Andrology Laboratory at MGH. Males between 20 and 54 years of age who were partners of subfertile couples were eligible; those who have had a vasectomy were excluded. Approximately 65% of eligible men agreed to participate. The most frequently cited reason for not participating was lack of time. A total of 406 men were recruited.

Personal Care Product Use Assessment

A trained research nurse administered a brief questionnaire to each subject at the time of his visit to the MGH andrology clinic for semen and urine sample collection. Information was obtained on personal care product use, smoking status, age, height, weight, race, and use of medications. Participants were specifically asked whether they had used hair gel/hair spray, lotion, aftershave, cologne, or deodorant in the 48 hr before the collection of the urine sample. They were also asked to record the time they last used the products within the 48-hr period.

Urinary Phthalate Monoester Measurement

A single spot urine sample was collected from each participant in a sterile plastic specimen cup (which was prescreened for phthalates) on the same day that the questionnaire was administered. The analytical approach has been described in detail [23] and adapted to both enable the detection of additional monoesters and improve efficiency of the analysis [24]. Briefly, measurement of monoester metabolites, namely, MEP, MBP, mono(2-ethylhexyl) phthalate (MEHP), MBzP, and monomethyl phthalate (MMP), entailed enzymatic deconjugation of the phthalates from their glucuronidated form, solid-phase extraction, HPLC separation, and tandem mass spectrometry detection. The limits of detection (LODs) were approximately 1 ng/mL. One method blank, two quality control samples (human urine spiked with phthalate monoesters), and two sets of standards were analyzed along with every 21 unknown urine samples. Analysts at the Centers for Disease Control and Prevention (CDC) were blind to all information concerning subjects. To control for urinary dilution, urinary concentrations were adjusted according to specific gravity. Specific gravity was measured using a handheld refractometer (National Instrument Company Inc., Baltimore, MD). The following formula was used to adjust phthalate concentrations by specific gravity: $Pc = P[(1.024 - 1)/SG - 1]$, where Pc represents specific gravity–corrected phthalate concentration (ng/mL), P is the measured phthalate concentration (ng/mL), and SG is the specific gravity of the sample. Specific gravity–adjusted monoester phthalate levels were used as continuous outcome variables in statistical models.

Statistical Methods

All analyses were performed using SAS software (version 8.1; SAS Institute Inc. Cary, NC). The use of each personal care product was categorized into a dichotomous variable (yes/no use in the 48 hr before the urine sample collection).

Because the phthalate monoester levels were not normally distributed, nonparametric tests were used to assess univariate associations between personal care product use and urinary phthalate levels. Multiple linear regression was used to explore the relationship between each of the five personal care products and each of the five log transformed monoester phthalate concentrations. In addition, a six-level sum variable was created, representing the number of different types of products used by a participant in the past 48 hr; possible values for this variable were 0, 1, 2, 3, 4, or 5. To determine if a dose–response relationship existed between urinary phthalate levels and the number of types of personal care products used, a trend test was performed using sum variable as an ordinal variable. For urinary phthalate concentrations that were below the LOD, a value equal to half the LOD was imputed (except when quantification was given) as follows: MEP, 0.605 ng/mL; MBzP, 0.235 ng/mL; MBP, 0.47 ng/mL; MEHP, 0.435 ng/mL; and MMP, 0.355 ng/mL.

After evaluating appropriateness using quadratic terms, we modeled age and body mass index (BMI; kilogram per square meter) as continuous independent variables. Smoking status was categorized as current smoker and current nonsmoker (includes ex-smokers and never smokers). Race was coded as African American, Hispanic, and other race, with Caucasian as the reference group. On the basis of biologic plausibility and statistical factors (i.e., change in parameter estimate), we included age, BMI, race, and smoking variables in all models as potential confounders.

To explore the relationship between time of product use and urinary levels of the phthalates, we regressed log-phthalate levels on the time between product use and urine sample collection (referred to as TIMEDIF). TIMEDIF was categorized into four intervals: product use 0–3 hr before urine sample collection (TIME0–3); > 3 but ≤6 hr (TIME3–6); > 6 but ≤8 hr (TIME6–8), and > 8 hr (TIME9). Approximately 75–85% of subject's product use was within 8 hr of urine collection, and therefore we used TIMEDIF > 8 hr as the reference category.

RESULTS

Subject Demographics

Of the 406 men recruited for an ongoing semen quality study, 37 did not provide urine samples. Of the remaining 369, specific-gravity analyses were not available for 32, leaving 338 for primary analysis. Additionally, one urine sample was missing MMP concentrations. The study population was composed largely of white (n = 275, 82%), nonsmoking men (n = 304, 91%) (Table 1). There were 19 African-American men, 18 Hispanic men, and 24 men of other race/ethnicity.

TABLE 1 Characteristics of Study Subjects (n = 338).

Characteristic	Value
Age [median (25%, 75%)]	35.0 (32.0, 39.1)
BMI [median (25%, 75%)]	27.5 (25.0, 30.6)
Racea [*n* (%)]	
White	275 (82)
Black/African American	18 (5)

TABLE 1 *(Continued)*

Characteristic	Value
Hispanic	19 (6)
Other	24 (7)
Smokingb [*n* (%)]	
Current smoker	31 (9)
Nonsmoker (ex- and never smoker)	304 (91)
Use of personal care products [*n* (%)]	
Lotionc	110 (34)
Hair gel/hair sprayd	121 (37)
Aftershavee	42 (13)
Deodorant[f]	299 (89)
Cologne[g]	94 (29)

[a]Race data missing for 2 men.
[b]Smoking data missing for 3 men.
[c]Lotion use data missing for 11 men.
[d]Hair gel/hair spray data missing for 7 men.
[e]Aftershave data missing for 8 men.
[f]Deodorant data missing for 1 man.
[g]Cologne data missing for 8 men.

Personal Care Product Use

Eleven men (3%) did not provide complete product use information (Table 1). Most men reported use of deodorant (89%), whereas fewer men reported using hair gel (37%), lotion (34%), cologne (29%), and aftershave (13%). Nine men (2.7%) did not use any of the personal care products listed on the questionnaire, 114 (33.7%) used only one type of product, 119 (35.3%) used two types of products, 71 (21%) used three types of products, 22 (6.5%) used four different types of products, and only 3 (0.9%) of the men used five or more different types of products within 48 hr of urine collection. The percentage of African-American (59%) and Hispanic (53%) men who reported using cologne within 48 hr of urine collection was higher the percentage of Caucasian men (25%) or men of other races (25%). Additionally, African-American men (65%) were more likely than Hispanic (44%), Caucasian (30%), or men of other races (43%) to use lotion. No other associations were seen between any other personal care products and race. Interestingly, men who used aftershave were almost twice as likely (18.5%) to also use cologne as non-aftershave users (9.8%) (chi squared p = 0.03). There were no consistent relationships among any of the other products used.

Urinary Phthalate Monoesters

There was a wide distribution of both specific gravity-adjusted (Table 2) and -unadjusted phthalate monoester levels (Table 3). Five phthalate monoesters were detected in 75–100% of subjects. MEP was the most prevalent (100%), followed by MBP (95%) and MBzP (90%). MEHP and MMP were both found in about 75% of subjects. Phthalate metabolite concentrations are presented both adjusted for specific gravity and unadjusted for comparison with other studies. The highest geometric mean levels were found for MEP (179 ng/mL), followed by MBP (16.6 ng/mL), MBzP (7.1 ng/mL), MEHP (6.6 ng/mL), and last, MMP (4.5 ng/mL).

Covariate Relationships

Race and cigarette smoking status were predictors of MEP and MBP levels (Table 4). We found significantly higher median MEP levels among African-American men (506 ng/mL) and Hispanic men (395 ng/mL) compared with Caucasian men (140 ng/mL) and those men categorized as other race (125 ng/mL). Median MBP levels in Caucasian men (15.3 ng/mL) were also lower than among African-American men (32.7 ng/mL) and Hispanic men (29.1 ng/mL), and among men identified as other race (26.5 ng/mL). Median MEP levels in current smokers (250 ng/mL) were significantly higher than among nonsmokers (143 ng/mL) (Table 4). BMI was weakly, although positively, correlated with MEP (Spearman correlation coefficient of 0.1, $p < 0.05$). Age was not associated with any of the five phthalate concentrations. Wilcoxon rank-sum tests showed positive associations between the sum variable for product use and African-American men and men of other races compared with Caucasians. BMI was positively associated with those identified as other race.

Product Use and Urinary Phthalate Relationship

In the univariate analyses, median MEP levels were higher among cologne users (265 ng/mL) compared with those who did not use cologne (108 ng/mL). Likewise, men who used aftershave had higher median MEP levels (266 ng/mL) than men who did not (133 ng/mL). Fragranced products such as cologne and aftershave contain relatively higher DEP levels than other personal care products. Figure 1, created on a subset of men who used cologne plus additional products, depicts the rise in MEP levels with specific combinations of personal care product use.

TABLE 2 Distribution of Specific Gravity–Adjusted Urinary Levels of Phthalate Monoesters: Percentiles and Summary Statistics (ng/mL).

		Percentile						
Phthalate	*n*	**5th**	**25th**	**50th**	**75th**	**95th**	**Mean ± SD**	**Geometric mean**
MEP	338	24.5	58.2	154	503	2, 030	490 ± 979	179
MEHP	338	< LOD	2.4	6.3	19.1	116.1	27.6 ± 69.1	6.6
MBP	338	3.1	10.3	16.5	30.6	68.2	76.2 ± 798.4	16.6
MBzP	338	< LOD	4.0	7.7	14.1	39.7	14.0 ± 34.6	7.1
MMP	337	< LOD	2.1	4.8	11.4	32.1	10.8 ± 22.8	4.5

LODs (ng/mL): MEP, 1.21; MBzP, 0.47; MBP, 0.94; MEHP, 0.87; MMP, 0.71.

TABLE 3 Distribution of Unadjusted Urinary Levels of Phthalate Monoesters: Percentiles and Summary Statistics (ng/mL).

		Percentile						
Phthalate	*n*	**5th**	**25th**	**50th**	**75th**	**95th**	**Mean ± SD**	**Geometric mean**
MEP	338	17.0	48.5	145	457	1, 953	485 ± 1, 008	164
MEHP	338	< LOD	1.9	5.2	18.4	134.6	25.6 ± 60.1	6.0
MBP	338	2.2	7.8	14.5	31.7	75.1	85.6 ± 932.8	14.9
MBzP	338	< LOD	2.9	6.8	14.1	41.3	13.90 ± 32.4	6.0
MMP	337	< LOD	1.7	4.5	10.1	31.3	11.0 ± 31.6	4.1

LODs (ng/mL): MEP, 1.21; MBzP, 0.47; MBP, 0.94; MEHP, 0.87; MMP, 0.71.

TABLE 4 Median (25th and 75th Percentiles) Urinary Levels of Phthalate Monoesters (ng/mL)a by Race and Smoking Status.

	MEP	**MEHP**	**MBP**	**MBzP**	**MMP**
Race					
Black	506* (294, 1, 134)	7.4 (3.9, 8.7)	32.7* (18.1, 42.5)	10.7 (6.8, 21.4)	6.0 (2.0, 12.0)
Hispanic	395* (83.3, 1, 076)	5.6 (3.3, 20.7)	29.1* (17.3, 42.4)	12.2 (3.5, 19.5)	5.2 (1.9, 11.9)
Other	125 (40.3, 218)	7.1 (1.7, 10.8)	26.5* (7.1, 38.4)	6.4 (2.3, 11.9)	4.6 (2.0, 11.3)
White	140 (56.6, 469)	6.2 (2.3, 20.7)	15.3 (9.9, 26.9)	7.4 (4.0, 14.2)	7.3 (3.3, 11.3)
Smoking					
Yes	250* (96.5, 826)	5.0 (1.8, 12.6)	20.9 (10.8, 46.9)	8.0 (4.3, 17.4)	8.5 (4.6, 18.2)
No	144 (57.5, 465)	6.4 (2.5, 19.6)	16.3 (10.3, 30.0)	7.7 (4.0, 14.2)	4.5 (2.1, 10.4)

[a]Specific gravity–adjusted phthalate levels.
*Univariate regression analysis $p \leq 0.05$; reference group for race is whites.

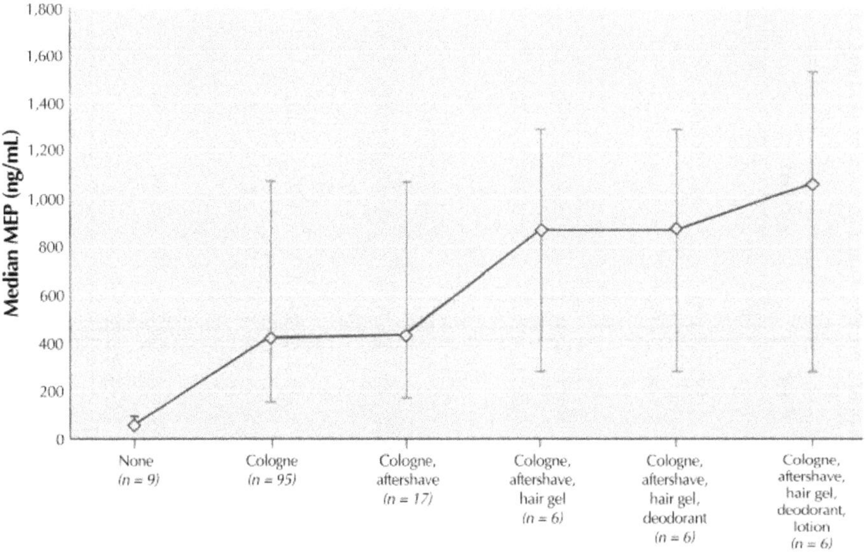

FIGURE 1 Specific-gravity–adjusted urinary MEP concentration according to combinations of product types used. Data points represent medians; error bars represent 25th and 75th percentiles.

Median MBP was lower among men who had used deodorant (16.3 ng/mL) compared with those who did not use deodorant (22.5 ng/mL). The use of lotion was associated with lower median levels of MBP (14.9 ng/mL), MBzP (6.1 ng/mL), and MEHP (4.4 ng/mL) compared with men who did not use lotion (MBP, 16.8 ng/mL; MBzP, 8.6 ng/mL; MEHP, 7.2 ng/mL) (Table 5).

Men of Hispanic, Caucasian, and other races who used cologne had considerably higher median MEP levels (981, 444, and 178 ng/mL, respectively) than non-cologne users of similar race (138, 102, and 116 ng/mL; $p = 0.09$, < 0.001, and 0.13, respectively). Interestingly, African-American men who used cologne had 30% lower median MEP levels compared with non-cologne users (371 ng/mL vs. 508 ng/mL), although the differences were not statistically significant. Hispanic and Caucasian men had substantially higher MEP levels if they used aftershave (1076 and 220 ng/mL) than if they did not (138 and 126 ng/mL; $p = 0.08$ and 0.03, respectively). African-American men who used aftershave had 33% lower MEP levels compared with non-aftershave users (340 ng/mL vs. 508 ng/mL), although the differences were not statistically significant. No other race/product associations were observed.

Multiple Linear Regression

In multiple linear regression models, after adjusting for race, smoking status, BMI, and age, urinary levels of MEP were 2.57 times higher among men who had used cologne and 1.71 times higher among aftershave users compared with men who did not report the use of these products ($p < 0.0001$ and 0.02, respectively) (Table 6). There was also a dose–response relationship between urinary phthalate MEP levels and the

TABLE 5 Median (25th and 75th Percentiles) Urinary Levels of Phthalate Monoesters (ng/mL)a by Personal Care Products Used 48 Hr Before Urine Sample Collection.

	MEP	MEHP	MBP	MBzP	MMP
Lotion					
Yes	136 (60.0, 438)	4.6* (1.6, 11.5)	15.7* (8.5, 28.1)	6.1* (3.0, 11.5)	4.2 (2.0, 10.4)
No	160 (56.6, 528)	7.2 (2.7, 20.4)	16.8 (10.3, 31.7)	8.6 (4.6, 15.1)	5.1 (2.4, 12.1)
Cologne					
Yes	422* (155, 1076)	5.3 (1.8, 16.9)	18.6 (12.2, 33.6)	10.5 (4.6, 16.7)	4.7 (2.6, 13.2)
No	110 (48.0, 293)	6.6 (2.5, 19.1)	15.4 (9.6, 29.2)	6.8 (3.9, 13.8)	4.9 (2.1, 10.3)
Deodorant					
Yes	165 (60.4, 534)	6.1 (2.4, 20.7)	16.3 (10.3, 29.2)	7.7 (4.0, 14.2)	4.7 (2.0, 11.3)
No	91.4 (36.4, 323)	6.3 (2.2, 13.7)	22.5 (10.8, 39.0)	7.5 (4.9, 14.0)	6.9 (2.6, 12.6)
Aftershave					
Yes	266* (123.2, 625)	6.1 (2.7, 13.8)	17.9 (10.9, 33.6)	7.9 (4.2, 16.2)	4.8 (2.4, 8.6)
No	135 (54.5, 477)	6.3 (2.3, 18.9)	16.3 (10.2, 30.2)	7.6 (3.9, 14.1)	4.8 (2.1, 12.0)
Hair gel/spray					
Yes	182 (57.8, 547)	7.7 (2.6, 21.6)	16.0 (8.9, 24.5)	8.0 (4.2, 14.2)	4.9 (2.0, 11.4)
No	139 (58.2, 464)	5.9 (2.1, 14.6)	16.7 (11.1, 31.7)	7.6 (4.0, 14.2)	4.8 (2.3, 11.7)

[a]Specific-gravity–adjusted phthalate levels.
*$p \leq 0.05$ in multivariate linear regression models adjusted for age, BMI, race, and smoking.

TABLE 6 Multiplicative Factorsa (95% Confidence Interval) for a Change in Urinary Phthalate Monoester Levelb Associated with Use of Personal Care Products Within the Past 48 Hr (N = 323).

Product type	MEP	MEHP	MBP	MBzP	MMP
Lotion	0.97 (0.70–1.33)	0.66 (0.44–0.99)	0.69 (0.53–0.88)	0.66 (0.50–0.87)	0.92 (0.66–1.29)
Cologne	2.57 (1.88–3.53)	0.96 (0.63–1.46)	1.06 (0.82–1.38)	1.16 (0.88–1.54)	1.19 (0.84–1.67)
Deodorant	1.24 (0.75–2.05)	1.23 (0.65–2.34)	0.70 (0.47–1.04)	0.95 (0.62–1.47)	0.94 (0.55–1.59)
Aftershave	1.71 (1.10–2.64)	0.93 (0.53–1.64)	1.00 (0.70–1.43)	1.16 (0.80–1.69)	0.97 (0.61–1.55)
Hair gel	1.15 (0.85–1.57)	1.23 (0.83–1.81)	0.92 (0.72–1.17)	0.95 (0.73–1.24)	0.99 (0.72–1.35)

[a]All models are adjusted for age, BMI, smoking, and race. Multiplicative factors represent multiplicative changes in phthalate levels associated with use of specific personal care products within the past 48 hr after back-transformation of phthalate concentrations: 1.0, no change in urinary phthalate level; < 1.0, multiplicative decrease in phthalate level; > 1.0, multiplicative increase in phthalate level.
[b]In all models, log transformations of specific gravity–adjusted phthalate concentrations were used.

number of types of personal care products used. For every additional type of product used, MEP concentrations increased 33% (95% confidence interval, 14–53%; trend test $p = 0.0002$) (Figure 2). The use of deodorant was associated with 30% lower MBP levels ($p = 0.08$). MBP, MBzP, and MEHP levels were 31% ($p = 0.004$), 34% ($p = 0.003$), and 34% ($p = 0.003$) lower, respectively, among men who had used lotion within the past 48 hr before urine collection compared with men who had not.

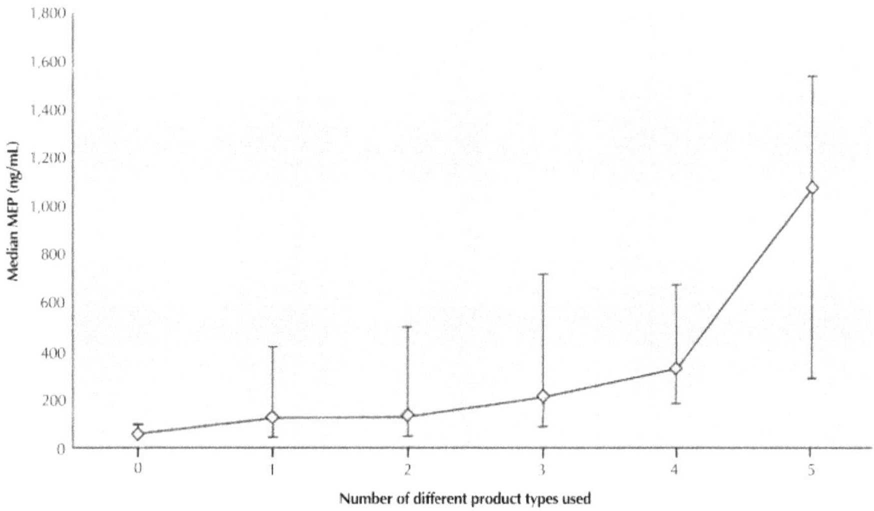

FIGURE 2 Specific-gravity–adjusted urinary MEP concentrations according to number of product types used. Data points represent medians; error bars represent 25th and 75th percentiles.

Time of Product Use

In secondary analyses, we explored the relationship between time of product use and urinary levels of phthalate monoesters. Statistical power was limited in these secondary analyses as a result of small sample sizes, generally fewer than 15 subjects for each of the four TIME strata. The analyses were performed only among users of each specific product. Cologne use at TIME0–3, TIME3–6, and TIME6–8 compared with cologne use at TIME9 was associated with an increase in MEP of 1.7-fold ($p = 0.17$), 2.8-fold ($p < 0.01$), and 1.1-fold ($p = 0.75$), respectively. No consistent time trends were observed for the other phthalates and cologne use. Aftershave was inconsistently associated with a 2- to 3-fold increase in MEP levels—3.0-fold increase at TIME0–3 ($p = 0.15$), 2.0-fold increase at TIME3–6 ($p = 0.25$), and 2.6-fold increase at TIME6–8 ($p = 0.11$)—compared with aftershave use at TIME9. No time trends were observed for the other phthalates and aftershave use. For lotion use at TIME0–3, TIME3–6, and TIME6–8, MBP concentration increased 1.9-fold ($p = 0.03$), 1.2-fold ($p = 0.55$), and 1.2-fold ($p = 0.52$) compared with lotion use at TIME9. No significant time relationships were found between lotion use and any other phthalate or between deodorant or hair gel use and any of the phthalates.

DISCUSSION

In the present study, men who used cologne and/or aftershave within the 48-hr period before the collection of the urine sample had higher urinary levels of MEP. This is not unexpected because previous studies have demonstrated that DEP, the parent compound of MEP, is an ingredient in many colognes, deodorants, and fragranced products [4, 5] and that percutaneous absorption of DEP occurs [1, 14, 20, 25]. More striking is the steepness of the dose–response relationship between the number of product types used in the 48 hr before urine collection and urinary MEP levels. DEP was found in 71% of the personal care products tested in one study [4], whereas DEHP, DBP, and BBzP were found in fewer than 10% of products. In another study, DEP was found in 57% of the perfumes and 25% of the deodorants surveyed; DBP, DEHP, and BBzP were not detected in any of the deodorants and in fewer than 27% of the perfumes [5]. Therefore, it is plausible that MEP would have a strong relationship with multiple product use, whereas the other phthalate monoesters would not.

Interestingly, the use of body lotion was associated with lower levels of MBP, MBzP, and MEHP. The reason for this relationship is not known, although several hypotheses are plausible. It is possible that other ingredients in body lotion may act as a barrier to the absorption of DBP, BBzP, and DEHP. It is also feasible that men who use lotion use fewer other personal care products. However, chi-squared tests did not show significant inverse relationships between the use of body lotion and other products (data not shown). An alternative explanation is that the urinary levels of these monoesters reflect exposure to their parent phthalates other than by use of personal care products. Percutaneous absorption after dermal exposure is expected to be lower for DBP, BBzP, and DEHP than for DEP.

The quantities of phthalates present in different brands of deodorant, aftershave, hair gel/hair spray, lotion, and cologne are known to be quite variable [4, 5]. In the present study, because information on the use of specific brand name products was not gathered, the analysis was performed by category of product. This approach is likely to introduce bias toward the null because not all products within a given category contain phthalates and those that do contain phthalates do so at variable concentrations. Because the participants in this study are all male, it is unclear whether the findings of this study may be generalizable to women, who may use different types and combinations of personal care products.

It is unclear why current smokers had higher levels of MEP. The results, however, were unstable because the sample size was small: only 31 men (9%) were current smokers. One potential explanation is that smoking may alter the toxicokinetics of DEP. Although DBP, unlike DEP, is listed as an ingredient in the filters of Phillip Morris cigarettes [26], MBP was not found to be related to current smoking status.

Racial differences in urinary levels of MEP and MBP were consistent with previous data from the National Health and Nutrition Examination Survey (NHANES) 1999–2000 that have shown African Americans and Hispanics have higher urinary levels of MEP and MBP than do Caucasians [27, 28]. In our study we explored the MEP and race associations for use of specific personal care products. The higher MEP levels for Hispanic than for Caucasian men appeared related to differentially higher cologne and aftershave use. Interestingly, the higher urinary MEP levels in African-

American than in Caucasian men did not appear to be related to higher cologne and/or aftershave use. Therefore, the use of other products not identified in this study, different sources of exposure to DEP, or differential toxicokinetics may be driving the high MEP levels among African-American men. After accounting for race, age, and smoking status in the statistical models, MEP levels were still significantly higher among cologne and aftershave users; African-American race remained an independent predictor of MEP levels. However, it is important to note that only 18 African Americans participated in the study, and these findings may be related to chance because of the small numbers. Further study on racial/ethnic differences is warranted.

In an earlier study on the relationship between demographic characteristics and urinary phthalate levels among a nonrepresentative subset of 289 participants of NHANES III, MBP, MBzP, and MEHP were higher in individuals of low socioeconomic status [29]. Urban residence was also significantly associated with higher MEHP and MBP levels. Socioeconomic status and area of residence were not controlled for in the present study, and these factors could potentially account for some of the differences measured between the racial groups. Finally, it is also possible that higher personal care product use or the selection of certain types of products among racial groups may contribute to differences in urinary phthalate levels.

The time elapsed between product use and urine sample collection influenced the relationship between cologne use and MEP concentrations. MEP was 2.7-fold higher when cologne was used between 3 and 6 hr before urine collection compared with when it was used 8 hr or more before urine collection. Therefore, to best assess the relationship of cologne use on urinary MEP levels, we suggest that urine collection should occur 3–6 hr after cologne use.

When time of lotion use was not accounted for in the analysis, there was an inverse association between urinary levels of MBP and lotion use. However, in analyses in which time of use was explored, MBP concentrations were significantly higher within the first 3 hr after lotion use compared with lotion use 8 hr or more before. The lotion use MBP relationship may require a larger data set to determine how use correlates with MBP levels in urine samples collected at variable times after applying lotion.

Although aftershave use between 0 and 8 hr before urine collection was associated with 2- to 3-fold higher MEP concentration compared with aftershave use more than 8 hr before urine collection, each strata had fewer than 10 subjects, and the reference group had only 15. This could explain why the aftershave–time of use relationships did not reach statistical significance.

To put these findings into perspective, a comparison with previous work is offered. The interquartile difference (443 ng/mL) in MEP, associated with increased DNA damage in sperm (Duty et al. 2003b), was approximately 2- to 3-fold higher than the difference in levels of MEP observed between men who did versus those who did not use cologne (312 ng/mL) or aftershave (131 ng/mL), respectively. MBP and MBzP, found in our previous study to be associated with decreased sperm motility and concentration [9], were not found to be associated with aftershave or cologne use.

CONCLUSIONS

Cologne and aftershave use were associated with significantly higher urinary MEP levels after controlling for age, BMI, smoking, and race. Additionally, a dose–response relationship was found between the number of different types of personal care products used and MEP urinary concentrations. Interestingly, lotion was inversely associated with most phthalate levels. Secondary analysis revealed that, for cologne, product use 3 to 6 hr before urine collection was most predictive of urinary MEP concentration. However, for lotion, product use in the 3 hr before urine collection was most predictive for MBP concentration.

The identification of personal care products as contributors to phthalate body burden is an important step in exposure characterization. Additionally, the results of this study suggest that the time that products are used in relation to the time that the urinary samples are collected should be documented. This will help reduce random measurement error in statistical analysis. Further work is needed to identify additional predictors of phthalate exposure.

KEYWORDS

- **Diethyl phthalate**
- **Monoethyl phthalate**
- **Personal care products**
- **Phthalates**
- **Urinary metabolites**

REFERENCES

1. ATSDR 1995. Toxicological profile for diethyl phthalate. Atlanta, GA: Agency for Toxic Substances and Disease Registry. Available: http://www.atsdr.cdc.gov/toxprofiles/tp73.html [accessed 23 May 2003].
2. ATSDR 2001. Toxicological profile for di-n-butyl phthalate. Atlanta, GA: Agency for Toxic Substances and Disease Registry. Available: http://www.atsdr.cdc.gov/toxprofiles/tp135.html [accessed 23 May 2003].
3. ATSDR 2003. Toxicological profile for di(2-ethylhexyl) phthalate. Atlanta, GA: Agency for Toxic Substances and Disease Registry. Available: http://www.atsdr.cdc.gov/toxprofiles/tp9.html [accessed 23 May 2003].
4. Houlihan J, Brody C, Schwan B. 2002. Not too pretty: phthalates, beauty products and the FDA. Environmental Working Group, Coming Clean, and Health Care Without Harm. Available: http://www.nottoopretty.org/images/NotTooPretty_final.pdf [accessed 3 September 2003].
5. Koo HJ, Lee BM. Estimated exposure to phthalates in cosmetics and risk assessment. Journal of Toxicology Environmental Health 2004; 67:1901–1914.
6. Li LH, Jester WF, Orth JM. Effects of relatively low levels of mono-(2-ethylhexyl) phthalate on cocultured Sertoli cells and gonocytes from neonatal rats. Toxicology of Applied Pharmacology 1998; 153(2):258–265.
7. Parks LG, Ostby JS, Lambright CR, Abbott BD, Klinefelter GR, Barlow NJ, et al. The plasticizer diethylhexyl phthalate induces malformations by decreasing fetal testosterone synthesis during sexual differentiation in the male rat. Toxicology Science 2000; 58(2):339–349.

8. Duty SM, Singh NP, Silva MJ, Barr DB, Brock JW, Ryan L, et al. The relationship between er vironmental exposures to phthalates and DNA damage in human sperm using the neutral com assay. Environmental Health Perspectives 2003a; 111:1164–1169.
9. Duty SM, Silva MJ, Barr DB, Brock JW, Ryan L, Chen Z, et al. Phthalate exposure and huma semen parameters. Epidemiology 2003b; 14(3):269–276.
10. Swan SH, Main KM, Liu F, Stewart SL, Kruse RL, Calafat AM, et al. Decrease in anogenit distance among male infants with prenatal phthalate exposure. Environmental Health Perspec tives 2005; 113(8):1056–61.
11. Howard J, Page N, Perkins M. 2001. Toxicology tutor II: Toxicokinetics, absorption: derm route. Bethesda, MD: Division of Specialized Information Services, National Library of Med cine, National Institutes of Health. Available: http://www.sis.nlm.nih.gov/enviro/ToxTutor/Tox a24.htm [accessed 7 November 2003].
12. Wester RC, Melendres J, Sedik L, Maibach H, Riviere JE. Percutaneous absorption of salicyli acid, theophylline, 2, 4-dimethylamine, diethyl hexyl phthalic acid, and p-aminobenzoic acid i the isolated perfused porcine skin flap compared to man in vivo. Toxicology of Applied Pharm cology 1998; 151(1):159–165.
13. Hagedorn-Leweke U, Lippold BC. Absorption of sun-screens and other compounds throug human skin in vivo: derivation of a method to predict maximum fluxes. Pharm Res 199 12(9):1354–1360.
14. Mint A, Hotchkiss SAM, Caldwell J. Percutaneous absorption of diethyl phthalate through r and human skin in vitro. Toxicol In Vitro. 1994; 8:251–256.
15. Barber ED, Teetsel NM, Kolberg KF, Guest D. A comparative study of the rates of in vitr percutaneous absorption of eight chemicals using rat and human skin. Fundamentals of Applie Toxicology 1992; 19(4):493–497.
16. Deisinger PJ, Perry LG, Guest D. In vivo percutaneous absorption of DEHP from DEHP plasticized polyvinyl chloride film in male Fischer 344 rats. Food Chemistry Toxicology 199 36(6):521–527.
17. Elsisi AE, Carter DE, Sipes IG. Dermal absorption of phthalate diesters in rats. Fundamentals c Applied Toxicology 1989; 12:70–77.
18. Melnick RL, Morrissey RE, Tomaszewski KE. Studies by the National Toxicology Program o di(2-ethylhexyl)phthalate. Toxicol Ind Health 1987; 3(2):99–118.
19. Ng KM, ChuI, Bronaugh RL, Franklin CA, Somers DA. Percutaneous absorption and metabc lism of pyrene, benzo[a]pyrene, and di(2-ethylhexyl) phthalate: comparison of in vitro and i vivo results in the hairless guinea pig. Toxicology of Applied Pharmacology 1992; 115(2):216 223.
20. Scott RC, Dugard PH, Ramsey JD, Rhodes C. In vitro absorption of some o-phthalate diester through human and rat skin. Environmental Health Perspectives 1987; 74:223–227.
21. Hauser R, Meeker J, Park S, Silva M, Calafat A. Temporal variability of urinary phthalate leve in men of reproductive age. Environmental Health Perspectives 2004; 112:1734–1740.
22. Hoppin JA, Brock JW, Davis BJ, Baird DD. Reproducibility of urinary phthalate metabolites i first morning urine samples. Environmental Health Perspectives 2002; 110:515–518.
23. Blount BC, Milgram KE, Silva NJ, Malek NA, Reidy JA, Needham LL, et al. Quantitative dete tion of eight phthalate metabolites in human urine using HPLC-APCI-MS/MS. Anal Chem 200 72(17):4127–4134.
24. Silva MJ, Malek NA, Hodge CC, Reidy JA, Kato K, Barr DB, et al. Improved quantitativ detection of 11 urinary phthalate metabolites in humans using liquid chromatography-atmc spheric pressure chemical ionization tandem mass spectrometry. Journal of Chromatogr B 200 789(2):393–404.
25. Api AM. Toxicological profile of diethyl phthalate: a vehicle for fragrance and cosmetic ingred ents. Food Chemistry Toxicology 2001; 39:97–108.
26. Phillip Morris. 2004. Ingredients in cigarettes. Richmond, VA: Phillip Morris. Available: http: www.philipmorrisusa.com/product_facts/ingredients/non_tobacco_ingredients.asp [accessed 2 November 2000].

27. CDC 2003. Second national report on human exposure to environmental chemicals. Atlanta, GA: Centers for Disease Control and Prevention. Available: http://www.cdc.gov/exposurereport/2nd/phthalates.htm [accessed 1 August 2003].
28. Silva MJ, Barr DB, Reidy JA, Malek NA, Hodge CC, Caudill SP, et al. Urinary levels of seven phthalate metabolites in the US population from the National Health and Nutrition Examination Survey (NHANES) 1999–2000. Environmental Health Perspectives 2004; 112:331–338.
29. Koo J-W, Parham F, Kohn MC, Masten SA, Brock JW, Needham LL, et al. The association between biomarker-based exposure estimates for phthalates and demographic factors in a human reference population. Environmental Health Perspectives 2002; 110:405–410.

10 Pesticides in House Dust

Lesliam Quirys-Alcal6, Asa Bradman, Marcia Nishioka, Martha E. Harnly, Alan Hubbard, Thomas E. McKone, Jeannette Ferber, and Brenda Eskenazi

CONTENTS

INTRODUCTION

Studies report that residential use of pesticides in low-income homes is common because of poor housing conditions and pest infestations; however, exposure data on contemporary-use pesticides in low-income households is limited. We conducted a study in low-income homes from urban and agricultural communities to: characterize and compare house dust levels of agricultural and residential-use pesticides; evaluate the correlation of pesticide concentrations in samples collected several days apart; examine whether concentrations of pesticides phased-out for residential uses, but still used in agriculture (i.e., chlorpyrifos and diazinon) have declined in homes in the agricultural community; and estimate resident children's pesticide exposures via inadvertent dust ingestion.

In 2006, we collected up to two dust samples 5–8 days apart from each of 13 urban homes in Oakland, California and 15 farmworker homes in Salinas, California, an agricultural community (54 samples total). We measured 22 insecticides including organophosphates (chlorpyrifos, diazinon, diazinon-oxon, malathion, methidathion, methyl parathion, phorate, and tetrachlorvinphos) and pyrethroids (allethrin-two isomers, bifenthrin, cypermethrin-four isomers, deltamethrin, esfenvalerate, imiprothrin, permethrin-two isomers, prallethrin, and sumithrin), one phthalate herbicide (chlorthal-dimethyl), one dicarboximide fungicide (iprodione), and one pesticide synergist (piperonyl butoxide).

More than half of the households reported applying pesticides indoors. Analytes frequently detected in both locations included chlorpyrifos, diazinon, permethrin, allethrin, cypermethrin, and piperonyl butoxide; no differences in concentrations or loadings were observed between locations for these analytes. Chlorthal-dimethyl was detected solely in farmworker homes, suggesting contamination due to regional agricultural use. Concentrations in samples collected 5–8 days apart in the same home were strongly correlated for the majority of the frequently detected analytes (Spearman $\rho = 0.70$-1.00, $p < 0.01$). Additionally, diazinon and chlorpyrifos concentrations in Salinas farmworker homes were 40–80% lower than concentrations reported in samples from Salinas farmworker homes studied between 2000–2002, suggesting a temporal reduction after their residential phase-out. Finally, estimated non-dietary pesticide intake for resident children did not exceed current U.S. Environmental Protection Agency's (U.S. EPA) recommended chronic reference doses (RfDs).

Low-income children are potentially exposed to a mixture of pesticides as a result of poorer housing quality. Historical or current pesticide use indoors is likely to contribute to ongoing exposures. Agricultural pesticide use may also contribute to additional exposures to some pesticides in rural areas. Although children's non-dietary intake did not exceed U.S. EPA RfDs for select pesticides, this does not ensure that children are free of any health risks as RfDs have their own limitations, and the children may be exposed indoors via other pathways. The frequent pesticide use reported and high detection of several home-use pesticides in house dust suggests that families would benefit from integrated pest management strategies to control pests and minimize current and future exposures.

BACKGROUND

Young children are particularly vulnerable to adverse health effects that may result from pesticide exposures. For example, in utero and/or postnatal chronic exposures to organophosphorous (OP) pesticides have been associated with poorer neurodevelopment in children [1–5], and altered fetal growth [6], and shortened gestational duration [7]. Animal studies have also shown that neonatal exposures to other contemporary use pesticides such as pyrethroids are associated with impaired brain development [8], changes in open-field behaviors, and increased oxidative stress [9].

Pesticides have been measured in residential environments, most notably in indoor dust [10–17]. Poor housing conditions in low-income homes, such as overcrowding and housing disrepair, are associated with pest infestations and increased home pesticide use in both urban and agricultural communities [18, 19], potentially increasing pesticide residues indoors. Additionally, the presence of farmworkers in the home and/or proximity of homes to nearby fields in agricultural communities have been associated with higher indoor pesticide concentrations [13, 20].

Several studies indicate that pesticide residues persist indoors due to the lack of sunlight, rain, temperature extremes, microbial action, and other factors that facilitate degradation [15]. Semi- and non-volatile pesticides (e.g., OPs and pyrethroids) have chemical properties that increase binding affinity for particles and the tendency to adsorb onto household surfaces such as carpet or dust, also prolonging their persistence indoors [11]. For example, pyrethroid pesticides have low vapor pressures, and high octanol/water (Kow) and water/organic carbon (Koc) partition coefficients which facilitate partitioning into lipids and organic matter and binding to particulate matter in dust [21]. Because of this, several studies suggest that house dust is an important pathway of pesticide exposure for children [11, 15, 17, 22]. Young children are particularly vulnerable to inadvertent ingestion of pesticide-contaminated dust due to their frequent hand-to-mouth activity and contact with indoor surfaces [15].

California (CA) has intense agricultural pesticide use [23], including OP insecticides. Due to their potential health effects in children, formulations of the OP insecticides, chlorpyrifos and diazinon, were voluntarily phased out for residential uses between 2001 and 2004 [24, 25]. One study showed that this residential phase-out resulted in decreased air concentrations among low-income households in New York City [26]. However, these OPs are still used in agriculture and trends in residential contamination of these compounds have not been studied in agricultural communities, where pesticide drift and transport from fields on work clothing may impact indoor pesticide concentrations [20]. It is also widely accepted that house dust is a reservoir for environmental contaminants with concentrations remaining fairly stable [11]; however, to our knowledge, only one study [27] has documented the temporal stability of pesticides in house dust focusing on the OP pesticide chlorpyrifos. Additionally, exposure data on other contemporary-use pesticides (e.g., pyrethroids) in low-income households is limited. In this study, we characterized and compared house dust levels of agricultural and residential-use pesticides from low-income homes in an urban community (Oakland, CA) and an agricultural community (Salinas, CA). We evaluated the correlation of several semi- and non-volatile pesticide concentrations in samples collected several days apart from the same general area in the home; and examined whether house dust concentrations of chlorpyrifos and diazinon declined in Salinas, CA after the U.S. Environmental Protection Agency's (EPA) voluntary residential phase-out of these compounds. Finally, we estimated resident children's potential non-dietary ingestion exposures to these indoor contaminants to determine if exposures via this pathway exceeded current U.S. EPA recommended guidelines.

METHODS

Study Population

Study participants included families with children between 3 and 6 years of age who were participating in a 16-day biomonitoring exposure study (to be presented elsewhere) conducted during July through September 2006. Through community health clinics and organizations serving low-income populations, we recruited a convenience sample of 20 families living in Oakland, CA, (an urban community in Alameda county) and 20 families living in Salinas, CA (an agricultural community with intense agricultural pesticide use in Monterey county). Participating families were Mexican American or Mexican immigrants and all Salinas households included at least one household member who worked in agriculture. The University of California, Berkeley Committee for the Protection of Human Subjects approved all study procedures and we obtained written informed consent from parents upon enrollment.

Data Collection

After obtaining informed consent from parents, bilingual staff administered a validated questionnaire [10] to ascertain demographic information on the children and household members, as well as information on factors potentially related to pesticide exposures such as: the presence of pest infestations and storage and use of pesticides in the previous "0–6 days," "7–30 days," "31–90 days," and " >90 days." We also conducted a home inspection to obtain information on housing quality and residential proximity to the nearest agricultural field or orchard. On dust collection days, parents were also asked if any pesticide applications had occurred in/around the home in the preceding 24 hours.

Dust Sample Collection

Using standard protocols [28], we collected dust samples from an area 1 to 2 m^2 with a High Volume Small Surface Sampler (HVS3) which collects particles >5 µm. Most dust samples were collected from carpets where parents indicated children spent time playing, except for two homes with no carpets or rugs, for which we collected samples from upholstered furniture using an attachment on the HVS3. To assess the consistency of concentrations within homes, we collected up to two dust samples, 5–8 days apart, from the same general location in each home. Dust samples were then manually sieved to obtain the fine fraction (<150 µm), which is more likely to adhere to human skin [15]. This fraction was stored at –80°C prior to shipment to Battelle Memorial Institute in Columbus, Ohio for laboratory analysis.

Laboratory Analysis

Of the 40 homes sampled, 15 Salinas farmworker and 13 Oakland urban homes had sufficient sample mass ($\geq$ 0.5 g) for analysis after measurement of other analytes (to be presented elsewhere). We analyzed two dust samples per home except for one home in each location from which one sample was analyzed, yielding a total of 54 dust samples. For this study, a total of 25 analytes were measured in every sample. Analytes measured included the OP insecticides chlorpyrifos, diazinon, malathion, methidathion, methyl parathion, phorate, tetrachlorvinphos, and one oxidation product of diazinon, diazinon-oxon; the pyrethroid insecticides bifenthrin, allethrin (two isomers), cypermethrin (four isomers), cis- and trans-permethrin, deltamethrin, esfenvalerate, imiprothrin, and prallethrin; the pesticide synergist commonly added to pyrethroid formulations piperonyl butoxide; the herbicide chlorthal-dimethyl; and the fungicide iprodione. We selected target analytes based on regional agricultural and non-agricultural use as reported in the California Department of Pesticide Regulation Pesticide Use Reporting database [29], active ingredients in pesticides used or stored indoors, detection in our prior studies [10, 13], and laboratory feasibility.

To measure analytes, we modified a previously published laboratory method [10, 13]. Briefly, 0.5 g dust aliquots were fortified with 250 ng of two surrogate recovery standards (SRSs)—fenchlorphos and 13C12-trans-permethrin—and extracted using ultrasonication in 1:1 hexane:acetone. We used solid phase extraction for sample clean-up, concentrated extracts to 1 mL and then fortified them with an internal standard, dibromobiphenyl. Concentrated extracts were analyzed with an electron impact gas chromatrography mass spectrometer in the multiple ion detection mode (Phenomenex ZB-35 column, 30 m × 0.25 mm ID, 0.25 μm film) with temperatures programmed from 130–340°C at 6°C/min. For each sample analysis set, we analyzed seven calibration curve solutions ranging from 2 to 750 ng/mL (five times higher for deltamethrin) and used a linear least squares regression and the internal method of quantification to prepare calibration curves. A solvent method blank, matrix spike sample (spike = 250 ng), and duplicate study sample were included in each sample analysis set for quality assurance and quality control purposes. We also determined the relative percent difference of the duplicate samples for each analyte measured to ensure that the analytical precision was within acceptable limits.

No analytes were detected in the four solvent method blanks, indicating no laboratory contamination. Analyte recoveries in four randomly-selected matrix spike samples averaged 117 ± 19% for OP analytes, 115 ± 16% for pyrethroid analytes, 82 ± 5% for chlorthal-dimethyl, 112 ± 14% for iprodione; and average SRS recoveries were 113 ± 6% and 128 ± 5% for fenchlorphos and 13C12-trans-permethrin, respectively. The average relative percent difference in concentration for the 12 analytes detected in duplicate samples was 14 ± 18% (n = 43 difference values spread across 12 analytes), indicating good analytical precision.

DATA ANALYSIS

We first summarized demographic characteristics and computed descriptive statistics for all analytes by location. For subsequent analyses, we focused on analytes frequently detected (i.e., detection frequencies, DF ≥ 50%). Concentrations below the limit of detection (LOD) were assigned a value of LOD/$\sqrt{2}$ [30] and results were considered significant at $p < 0.05$.

We used Fisher's Exact tests to determine if analyte detection frequencies differed between locations. To assess differences in concentrations between study locations, we used linear regression models with a generalized estimating equations (GEE) approach [31] in order to report robust inference that accounts for the non-independence of repeated samples within households. Given the limited number of homes sampled and the homogeneity of the study population, we excluded demographic characteristics as covariates in GEE models. We also examined location differences using analyte loadings, ng/m2 [21]. We calculated loadings by multiplying analyte concentrations by the sieved fine mass and dividing by the area sampled.

To determine the correlation of analyte concentrations between the first and second collections, we computed Spearman rank-order correlations.

To examine temporal trends of chlorpyrifos and diazinon concentrations in farmworker homes after the residential phase-out, we used Wilcoxon Mann-Whitney tests to compare dust concentrations in the 15 (n = 29 samples) Salinas farmworker households sampled in 2006 from our present study with dust concentrations from a subset of 82 Salinas farmworker homes of participants in the CHAMACOS study [13] sampled between 2000 and 2002 (2000, n = 33; 2001, n = 36; 2002, n = 13), and 20 similar households sampled by Bradman et al. [10] in 2002. The same laboratory (Battelle Memorial Institute) and collection methods were used in all studies. In addition, we restricted comparisons to those study homes located in the same zip codes as the homes in the present study. If multiple dust samples were available from any of the study homes in the same year, including the present study, the mean analyte dust concentration was used in our analyses. There were no demographic or household differences between our previous studies and the present study; e.g., all households had at least one farmworker residing in the home and study participants generally represented the farmworker population in Salinas Valley: primarily Mexican or of Mexican descent; Spanish-speaking; low literacy; low income; and frequently reported pesticide applications in the home and wearing work clothes and shoes indoors. Homes were also located >200 feet from the nearest agricultural field. Using the California Department of Pesticide Regulation Pesticide Use Reporting (PUR) database [29], we also computed county-level agricultural and non-agricultural usage of these OP pesticides during 1999–2007 to determine whether temporal changes in residential dust concentrations were concurrent with regional use patterns. Non-agricultural uses included applications for landscape maintenance, public health, commodity fumigation, rights-of-way, and structural pest control applications by licensed applicators, which are reported to the state.

Finally, to determine if exposures via the non-dietary ingestion pathway exceed U.S. EPA guidelines for the children in the present study, we calculated hazard quotients (HQ) for the majority of the detected analytes. We focused on the children given their unique vulnerabilities to environmental toxicants [32]. We calculated the HQ as the ratio of the child's potential daily toxicant intake at home via non-dietary ingestion (mg/kg/day) to the specific toxicant chronic reference dose, RfD, (mg/kg/day). The potential daily toxicant intake was calculated as follows:

$$PDI(mg/kg/day) = (Cdust \times IR)/BWchild$$

where Cdust is the analyte dust concentration in the child's home (mg/g), IR is the dust intake rate—assumed to be 0.10 g/day (100 mg/day) [33], and BWchild is the child's body weight (kg) obtained at the initial visit. We used chronic RfDs because children ingest small amounts of dust every day [33]. Chronic oral RfDs were available for 14 of the detected pesticides. For those pesticides that have been re-registered in response to the Food Quality Protection Act [34], chronic population adjusted doses (cPAD) were used as the reference dose. An HQ >1.0 would suggest that the child's exposure via non-dietary ingestion, independent of other exposure routes, may exceed the U.S. EPA's RfD.

All statistical analyses were performed using Stata 10 for Windows (StataCorp, College Station, TX).

RESULTS

Household Demographics and Pesticide Use

Except for farmworker status, demographic characteristics were similar in both study locations (Table 1). Participating households were within 200% of the poverty line and approximately 50% or more of the homes had at least six household members. Although not statistically significant, pest sightings were more commonly reported in Oakland urban homes compared to Salinas farmworker homes. Most participants reported using pesticides indoors in the three months preceding the study (67% and 85% of farmworker and urban homes, respectively) and the most common location of use was the kitchen. Hand-held pyrethroid sprays were the most common formulation and application method in both locations; applications were mostly targeted at ants and cockroaches. Participants from three homes (one Salinas farmworker home and two Oakland urban homes) reported applying pyrethroid insecticides between the two sampling dates. No products with OP insecticides were stored or reported applied in the homes, at the workplace or on pets. Most participants from Salina's households (80%) reported that farmworkers residing in the home wore their work clothing indoors and about half of them also wore their work shoes indoors. Approximately 27% of Salinas farmworker homes were located <1/4 mile from the nearest agricultural field or orchard.

Dust Levels: Trends and Location Differences

We detected 21 of the 25 analytes measured (Table 2). The majority of homes (93%) had at least three analytes detected in dust; 79% of the homes (n = 22) had at least six analytes detected and <1% (n = 2) of Salinas farmworker homes had up to 14 analytes detected in one sample. Cis- and trans-permethrin were the only insecticides detected in every home. Commonly detected OP pesticides included diazinon and chlorpyrifos. Diazinon was detected in 79% and 52% of the samples collected from Salinas farmworker and Oakland urban homes, respectively. Chlorpyrifos was detected in 5% and 36% of the samples collected from Salinas farmworker homes and Oakland urban homes, respectively. Other commonly detected analytes in samples collected from both locations included: allethrin (DF ≥ 80%), cypermethrin (DF ≥ 55%), and piperonyl butoxide (DF ≥ 86%). Detection frequencies were only significantly different between locations for chorthal-dimethyl, which was detected solely in Salinas farmworker homes.

TABLE 1 Demographic and Household Characteristics for Study Participants from Farmworker Homes in Salinas, CA, and Urban Homes in Oakland, CA.[a]

	Salinas farmworker homes (n = 15)		Oakland urban homes (n = 13)	
	n	(%)	n	(%)
Maternal education (highest grade completed)				
< completed 9th grade or lower	8	(53.3)	8	(61.5)
Grades 10–12 (no diploma)	3	(20.0)	1	(7.7)
High school diploma/GED or technical school	2	(13.3)	4	(30.8)
College graduate	2	(13.3)		---
Paternal education (highest grade completed)				

TABLE 1 *(Continued)*

	Salinas farmworker homes (n = 15)		**Oakland urban homes (n = 13)**	
	n	**(%)**	**n**	**(%)**
< completed 9th grade or lower	12	(85.7)	10	(83.3)
Grades 10–12 (no diploma)	1	(7.1)	1	(8.3)
High school diploma/GED or technical school	1	(7.1)	1	(8.3)
College graduate		---		---
Family income relative to federal poverty level[b]				
≤ Poverty level	10	(66.7)	9	(69.2)

TABLE 1 *(Continued)*

	Salinas farmworker homes (n = 15)		Oakland urban homes (n = 13)	
	n	(%)	n	(%)
> Poverty level but <200% poverty level	5	(33.3)	4	(30.8)
Number of household members				
3-5	8	(53.3)	5	(38.5)
> 6	7	(46.7)	8	(61.5)
Reported rodent sighting in the home				
Yes	2	(13.3)	3	(23.1)
No	13	(86.7)	10	(76.9)

	Salinas farmworker homes (n = 15)		Oakland urban homes (n = 13)	
	n	**(%)**	**n**	**(%)**
Reported cockroach sighting in the home				
Yes	3	(20.0)	5	(38.5)
No	12	(80.0)	8	(61.5)
Reported pesticide application in the last 3 months				
Yes	10	(66.7)	11	(84.6)
No	5	(33.3)	2	(15.4)
Farmworkers wore work clothing indoors[c]				

TABLE 1 *(Continued)*

	Salinas farmworker homes (n = 15)		Oakland urban homes (n = 13)	
	n	**(%)**	**n**	**(%)**
Yes	12	(80.0)		---
No	2	(20.0)		
Farmworkers wore work shoes indoors[c]				
Yes	7	(50.0)		---
No	7	(50.0)		
Farmworkers living in the home (past 3 months)				
0		---	11	(84.6)[d]

TABLE 1 (Continued)

	Salinas farmworker homes (n = 15)		Oakland urban homes (n = 13)	
	n	(%)	n	(%)
1-3	15	(100.0)	2	(15.4)
Farmworkers currently living in the home				
0	1	(6.7)		---
1-3	11	(73.3)		---
4-7	3	(20.0)		---
Distance of home to nearest field/orchard				
50-20 feet	1	(6.7)		---

TABLE 1 *(Continued)*

	Salinas farmworker homes (n = 15)		Oakland urban homes (n = 13)	
	n	(%)	n	(%)
> 200 feet-1/4 mile	3	(20.0)	---	
> 1/4 mile	11	(73.3)	---	

a. No statistically significant differences were observed between locations for demographic factors unrelated to farmworker status. b. Families' poverty levels were based on U.S. Department of Health and Human Services thresholds for 2006. Source: http://aspe.hhs.gov/POVERTY/06poverty.shtml. c. One participant in the Salinas group reported that the father was a farmworker during the eligibility screening; however, the father was not living in the home during the sample collection period so information is only available for 14 of the 15 farmworker households for this demographic characteristic. d. Two participants reported having a parent or parent's sibling working in a field/golf course doing maintenance/landscaping work potentially involving pesticide use; however, they were not doing this work during sample collection.

TABLE 2 Limits of Detection and Summary Statistics for Pesticide Dust Concentrations (Ng/G) in Samples Collected in 2006 from Low-Income Farmworker and Urban Homesa, b.

	Salinas farmworker homes (n = 29 samples collected from 15 homes)								Oakland urban homes (n = 25 samples collected from 13 homes)						
	LOD (ng/g)	DF	min	p25	p50	p75	p95	max	DF	min	p25	p50	p75	p95	max
Organophosphates															
Diazinon	4	79	--	8.21	14.4	18	35.8	56.4	52	--	--	6.98	18.1	133	139
Chlorpyrifos	10	55	--	--	21.9	28	135	200	36	--	--	--	34.9	43.7	56.4
Malathion	10	7	--	--	--	--	52.2	70.8	12	--	--	--	--	877	1160
Tetrachlorvinphos	50	10	--	--	--	--	252	271	4	--	--	--	--	--	15.8
Diazinon-oxon	4	ND	--	--	--	--	--	--	4	--	--	--	--	--	4.73

TABLE 2 *(Continued)*

	Salinas farmworker homes (n = 29 samples collected from 15 homes)								**Oakland urban homes (n = 25 samples collected from 13 homes)**						
	LOD (ng/g)	**DF**	**min**	**p25**	**p50**	**p75**	**p95**	**max**	**DF**	**min**	**p25**	**p50**	**p75**	**p95**	**max**
Methidathion	10	ND	--	--	--	--	--	--	ND	--	--	--	--	--	--
Methyl Parathion	10	ND	--	--	--	--	--	--	ND	--	--	--	--	--	--
Phorate	10	ND	--	--	--	--	--	--	ND	--	--	--	--	--	--
Pyrethroids															
cis-permethrin	4	100	45.9	84.9	568	908	5930[c]	6300[c]	100	11.6	84.4	291	946	21600	26700
trans-permethrin	4	100	88.4	144	952	1380	9170[c]	9690[c]	100	18.4	166	504	1620	36400	46800

TABLE 2 (Continued)

	Salinas farmworker homes (n = 29 samples collected from 15 homes)								Oakland urban homes (n = 25 samples collected from 13 homes)						
	LOD (ng/g)	DF	min	p25	p50	p75	p95	max	DF	min	p25	p50	p75	p95	max
Allethrin[d]	10	83	--	18.4	57.1	129	652[c]	694	80	--	20.376	50.5	158	276	289
Cypermethrin[e]	20	55	--	--	230	918	4540	13500	64	--	--	587	1050	5990	13100
Bifenthrin	10	14	--	--	--	--	23.8	23.9	44	--	--	--	45	2050	2120
Sumithrin	10	24	--	--	--	--	591	807	8	--	--	--	--	104	116
Deltamethrin	250	17	--	--	--	--	3780	5590	12	--	--	--	--	13000	16300
Imiprothrin	50	7	--	--	--	--	253	2140	4	--	--	--	--	--	160
Prallethrin	2	ND	--	--	--	--	--	--	4	--	--	--	--	--	33.6

TABLE 2 *(Continued)*

	Salinas farmworker homes (n = 29 samples collected from 15 homes)								Oakland urban homes (n = 25 samples collected from 13 homes)						
	LOD (ng/g)	**DF**	**min**	**p25**	**p50**	**p75**	**p95**	**max**	**DF**	**min**	**p25**	**p50**	**p75**	**p95**	**max**
Esfenvalerate	50	3	--	--	--	--	--	66.5	ND	--	--	--	--	--	--
Other															
Piperonyl butoxide[f]	2	86	--	30.9	92.3	283	9060	9350	96	--	51.6	353	751	40300	46600
Chlorthal-dimethyl[g]	2	97	--	13.3	16.3	23.5	34.1	34.8	ND	--	--	--	--	--	--
Iprodione[h]	100	ND	--	--	--	--	--	--	ND	--	--	--	--	--	--

a. Two samples were obtained from each home in both locations except for one home in each location due to inadequate sample volume. b. Samples were collected from carpets or area rugs with the exception of three samples from two farmworker homes, which were collected from furniture due to the absence of a carpet. c. Denotes that the reported concentration was observed in a furniture sample. d. Reported as the sum of two isomers (cis/trans) isomers. e. Reported as the sum of four isomers. f. Insecticide synergist. g. Phthalate herbicide h. Dicarboximide fungicide. Abbreviations and notation: LOD = limit of detection; DF = detection frequency (based on the number of samples obtained); ND or '--' indicates that analyte was not detected or detected <LOD so a summary statistic is not reported.

Median concentrations of diazinon, chlorpyrifos, permethrins, allethrin, and chlorthal-dimethyl were higher in Salinas farmworker homes compared to Oakland urban homes; however, only chlorthal-dimethyl concentrations were significantly different between locations. Analyses of location differences using pesticide loadings (ng/m^2) did not change our findings.

Dust concentrations from furniture samples in two farmworker homes were comparable to those collected from carpets in other farmworker homes for frequently detected OPs (diazinon and chlorpyrifos), piperonyl butoxide, and chlorthal-dimethyl, while for frequently detected pyrethroids, concentrations were generally at the upper end of the distribution. We observed the same general pattern when using loadings. Maximum permethrin concentrations in farmworker homes were observed in furniture samples; however, the highest permethrin concentrations were observed in carpet samples from urban homes. The highest loading observed for cypermethrin was collected from a furniture sample; however, higher loadings were observed in carpet samples from urban homes. No location differences in pesticide concentrations or loadings were observed when we excluded furniture samples from our analysis.

Some of the less frequently detected analytes (e.g., tetrachlorvinphos, sumithrin) were detected with greater frequency in Salinas farmworker homes and at higher maximum concentrations than in Oakland urban homes. Conversely, the 95th percentile and maximum concentrations for malathion and deltamethrin were higher among Oakland urban homes (Table 2).

Although not statistically significant, we generally observed higher dust concentrations in homes that reported recent pesticide use (i.e., within the last three months preceding the study) when pesticide containers were available to confirm the active ingredients. For example, in one home where bifenthrin had been applied less than a week before the first sample collection, concentrations were up to 200 times higher than the median concentration observed in other homes. Cypermethin was applied in one farmworker home, while imiprothrin was applied in two urban homes between the two sampling dates. For the farmworker home, cypermethrin dust concentrations were at the upper end of the distribution among other farmworker homes (between the 75th and 95th percentile concentrations reported). Imiprothrin was only detected in one of the urban homes that reported usage during the study; no other urban home had detectable imiprothrin levels indoors even though some of these households reported applying imiprothrin indoors prior to the study.

Concentrations in samples collected 5–8 days apart in the same home were positively and significantly correlated for the most frequently detected analytes (i.e., DF ≥ 0%), except allethrin; Spearman rank-order correlation coefficients ranged from 0.70 to 1.00 ($p < 0.01$) (Table 3).

TABLE 3 Spearman Rank-Order Correlation Coefficients for Dust Concentrations Between the First and Second Collections for the Most Frequently Detected Analytes[a].

Analyte	Salinas farmworker homes (n = 14)[b]	Oakland urban homes (n = 12)[b]	All homes (n = 26)[c]
		Spearman rho	
Organophosphates			
Diazinon	0.88**	0.97**	0.92*
Chlorpyrifos	0.83**	--	--
Pyrethroids			
cis-permethrin	0.78**	1.00**	0.91**
trans-permethrin	0.70**	1.00**	0.90**
Allethrin	0.49	0.18	0.36
Cypermethrin	0.87**	0.89**	0.89**
Synergist Ingredient			
Piperonyl butoxide	0.77*	0.97**	0.89**
Phthalate Herbicide			
Chlorthal-dimethyl	0.78*	--	--

a. Only those homes for which we were able to measure analytes in both dust samples wer included in these analyses. b. Spearman correlation coefficients are only provided when analy DF≥50% in respective locations at each collection. c. Correlation coefficients for all homes a provided when analyte DF≥50% in both locations and collections. ** $p \leq 0.0001$, * $p < 0.01$

Temporal Trends of Chlorpyrifos and Diazinon Dust Concentrations in Salina Farmworker Homes after the U.S. EPA's Residential Phase-Out

As noted earlier, residential formulations of chlorpyrifos and diazinon were volur tarily phased-out by the end of 2001 and between 2002 and 2004 [24, 25], respectivel

However, agricultural use of chlorpyrifos and diazinon in Monterey County generally increased from 1999–2007 (trendline in Figure 1), most notably for diazinon. Non-agricultural uses in Monterey County (i.e., applications for landscape maintenance,

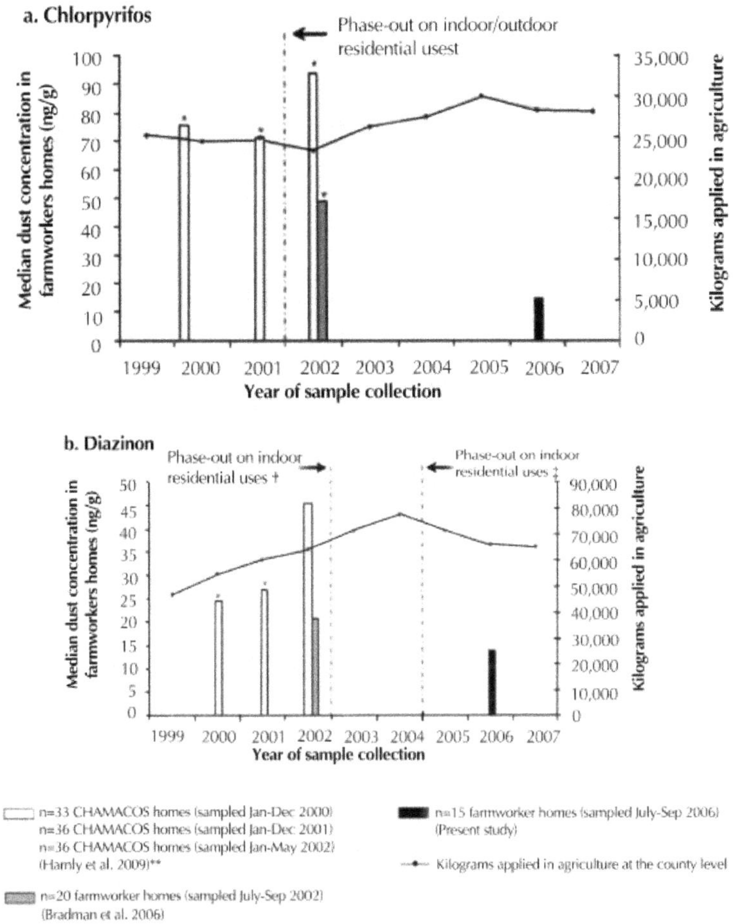

FIGURE 1 Median chlorpyrifos and diazinon dust concentrations in samples from farmworker **homes in** the city of Salinas, CA by year of collection and kilograms applied (trendline) at the county-level (Monterey County) for agricultural purposes from 1999-2007. † In December 2001 and 2002, residential products containing chlorpyrifos and diazinon, respectively were canceled. ‡Technical registrants were to buy back existing products from retailers by the end of December 2004. * Indicates that study had significantly higher dust concentrations compared to those observed in farmworker homes sampled in the present study (Wilcoxon Mann-Whitney tests, $p < 0.05$). ** CHAMACOS refers to the Center for the Health Assessment of Mothers and Children of Salinas longitudinal birth cohort study (Harnly et al. 2009).

public health, commodity fumigation, rights-of-way, and structural pest control) for both of these OP pesticides was a small fraction (<6%, ≈ 20-1, 500 kgs/yr) of agricultural use between 1999 and 2005, and declined further through 2006–2007 (Table 4). As shown in Figure 1, median dust concentrations of chlorpyrifos and diazinon were 70–80% and 40–50% lower, respectively, in the present farmworker homes sampled in 2006 compared to samples collected between 2000 and 2002 from farmworker homes in the same Salinas zip codes. Chlorpyrifos dust concentrations differed significantly between the present study and each of the previous studies (Wilcoxon Mann-Whitney tests, $p < 0.05$). Diazinon concentrations were significantly lower in the present study compared to CHAMACOS households sampled prior to 2002 ($p < 0.05$).

TABLE 4 Kilograms of Diazinon and Chlorpyrifos Applied in Monterey County from 1999–2007 for Non-Agricultural Applications[a, b].

Year	Diazinon (kgs)	Chlorpyrifos (kgs)
1999	717	1519
2000	841	678
2001	1301	355
2002	1094	760
2003	1076	101
2004	217	18
2005	96	54
2006	6	3
2007	2	<0.5

a. Pounds applied at the county level are reported by year by the California Depart ment of Pesticide Regulation in their Pesticide Use Reporting (PUR) database. Pound reported in the database were converted to kilograms. b. Non-agricultural application refer to uses such as landscape maintenance, public health, commodity fumigation rights-of-way, and structural pest control applications by licensed applicators. Source California Department of Pesticide Regulation Pesticide Use Regulation Database Available at: http://www.cdpr.ca.gov/docs/pur/purmain.htm

Estimated Non-Dietary Ingestion Intake and Hazard Quotients

For the 14 detected pesticides with available RfDs, none of the hazard quotients for resident children's non-dietary dust ingestion exceeded 1.0 (i.e., estimated intake did not exceed the RfD, Table 5).

DISCUSSION

We detected several pesticides in most homes, including OP pesticides previously phased-out for residential uses, pyrethroids, and the pesticide synergist piperonyl butoxide (PBO). Biological exposure metrics for these pesticides are relatively transient and highly variable, typically reflecting recent exposures [35]. However, consistent with other studies [15, 36], we found that dust serves as a stable matrix and indicator of potential indoor exposure for some pesticides. The high correlations observed in dust concentrations from samples collected 5–8 days apart suggests that, for some pesticides, measurements in house dust may be relatively stable indicators of potential indoor exposure over this time frame. To our knowledge, this is the first study to evaluate the correlation of concentrations within homes for several pesticides over a short sampling period.

Although the detection frequency for chlorpyrifos and diazinon was higher in Salinas than Oakland, we did not observe statistically significant differences in pesticide concentrations or loadings between locations. This is notable given that >28, 000 and 65, 000 kgs of chlorpyrifos and diazinon, respectively, were applied for agricultural purposes in Monterey County in 2006 and minimal applications (65 kgs and 3 kgs of chlorpyrifos and diazinon, respectively) occurred in Alameda County. Previously, we showed a significant correlation with local agricultural use and chlorpyrifos dust concentrations (but not diazinon) for homes throughout the Salinas Valley [13]. Mapping of dust concentrations and agricultural use suggests that chlorpyrifos dust concentrations are higher in the center of the Valley (south of the city of Salinas), where agricultural use is higher [13]. Farmworker homes in the present study were from the city of Salinas where the impact of drift from agricultural applications may have been lower. Additionally, our small sample size may have prevented us from observing significant differences in concentrations between locations for these OP pesticides as well as other analytes.

Malathion was not frequently detected in homes from either location; however, higher levels were observed in urban homes. This pesticide is used in agriculture and is also registered for use in home gardens, as a building perimeter treatment, as a wide-area spray for mosquitoes, and by prescription for head-lice control [37]. However, no parents reported treating their children for lice or using it themselves in their gardens. The main county uses for this OP pesticide in 2006 in the urban region were landscape maintenance and structural pest control [38]. These applications were reported more than 25 km away from the nearest study home, thus it is not readily apparent why higher levels were observed in urban homes although it should be noted that we only sampled a small number of homes.

We generally observed significantly lower house dust concentrations of chlorpyrifos and diazinon in the present study compared to levels measured in dust from homes located in the same zip codes sampled between 2000 and 2002 [10, 13], suggesting

TABLE 5 Summary Statistics on the Estimated Intake and Hazard Quotients (HQ) for All Study Children[a].

					Select Summary Statistics for HQs based on 54 dust samples from farmworker and urban children			
Analyte	RfD (mg/kg/dy)[b]	# samples with concentration >LOD	Range of Intake (mg/kg/day)[c]		p50	p75	p95	Max
Organophosphates			Min	Max				
diazinon	0.0002	36	--	7.0×10^{-07}	2.5×10^{-04}	4.3×10^{-04}	1.6×10^{-03}	3.5×10^{-03}
chlorpyrifos	0.00003	25	--	1.1×10^{-06}	--	4.9×10^{-03}	2.3×10^{-02}	3.8×10^{-02}
malathion	0.07	5	--	4.9×10^{-06}	--	--	2.8×10^{-06}	7.0×10^{-05}
tetrachlorvinphos	0.04	4	--	1.5×10^{-06}	--	--	2.2×10^{-05}	3.9×10^{-05}

					Select Summary Statistics for HQs based on 54 dust samples from farmworker and urban children			
Pyrethroids								
cis-permethrin[d]	0.25	54	5.4×10^{-08}	1.3×10^{-04}	9.7×10^{-06}	2.0×10^{-05}	8.4×10^{-05}	5.1×10^{-04}
trans-permethrin[d]	0.25	54	8.6×10^{-08}	2.2×10^{-04}	1.7×10^{-05}	3.2×10^{-05}	1.6×10^{-04}	8.9×10^{-04}
cypermethrin	0.06	32	--	6.4×10^{-05}	2.5×10^{-05}	8.4×10^{-05}	4.8×10^{-04}	1.1×10^{-03}
bifenthrin	0.015	15	--	1.1×10^{-05}	--	4.4×10^{-06}	2.6×10^{-05}	7.5×10^{-04}
sumithrin	0.0007	9	--	3.7×10^{-06}	--	--	3.2×10^{-03}	5.2×10^{-03}
deltamethrin	0.0033	8	--	9.0×10^{-05}	--	--	6.7×10^{-03}	2.7×10^{-02}
prallethrin	0.025	1	--	1.9×10^{-07}	--	--	--	7.4×10^{-06}

TABLE 5 *(Continued)*

					Select Summary Statistics for HQs based on 54 dust samples from farmworker and urban children			
esfenvalerate	0.02	1	--	4.4×10^{-07}	--	--	--	2.2×10^{-05}
Others								
chlorthal-dimethyl	0.01	28	--	1.8×10^{-07}	3.0×10^{-06}	7.2×10^{-06}	1.5×10^{-05}	1.8×10^{-05}
piperonyl butoxide	0.16	49	--	2.2×10^{-04}	5.0×10^{-06}	1.7×10^{-05}	2.5×10^{-04}	1.4×10^{-03}

a. A hazard quotient (HQ) was calculated as the ratio of the potential dust intake to the respective analyte reference dose (RfD). The HQ was only calculated for those analytes for which an RfD was available. b. Chronic population adjusted doses (cPADs) were used as the reference dose for chlorpyrifos (cPAD for children and females 13–50 years of age) and deltamethrin. Sources: IRIS database http://cfpub.epa.gov/ncea/iris/index.cfm?fuseaction=iris.showSubstanceList and EPA's Pesticide Registration Status: http://www.epa.gov/opp00001/reregistration/status.htm. c. Intake was calculated by multiplying the toxicant dust concentration by an ingestion rate of 0.10 g/day (100 mg/day) then dividing by the child-specific body weight (kg). d. The RfD available for "permethrin" was used for each individual isomer in our calculations. Notation: "—" value is not reported since dust concentrations were less than the limit of detection.

that indoor concentrations in the city of Salinas are decreasing despite continued agricultural use in the area. In New York City, air concentrations for these OP pesticides in low-income homes also significantly decreased between 2001 and 2004 [26]. The temporal declines in indoor concentrations reported here and in the New York City study may reflect the decreasing usage of these OP pesticides for home or structural applications per the U.S. EPA's residential phase-out. Nonetheless, despite declining concentrations indoors, detection of these OP pesticides, especially in Oakland where there was little agricultural or structural use, underscores their persistence indoors.

Compared to other studies in farmworker populations (Table 6), we observed lower median concentrations for chlorpyrifos [10, 13, 17, 22, 39, 40] and diazinon [10, 13, 40]. These farmworker studies generally reported a wider range of concentrations for these two OP pesticides and collected dust samples prior to the residential phase-out. One study by Curl et al. [22] reported a wider range of diazinon concentrations, but comparable median concentrations (10 ng/g). Although malathion was not frequently detected in our farmworker homes, a wider range of concentrations was reported in previous farmworker studies (Table 6) [10, 22, 40]. To our knowledge, only one other study has reported OP pesticide concentrations in low-income urban homes [41]. This study reported higher median concentrations for chlorpyrifos and diazinon in low-income urban housing units in Boston, MA. Homes in this study were sampled just after or during the residential phase-out of chlorpyrifos and diazinon, respectively (between July 2002 and August 2003).

Pyrethroids were detected in house dust in several study homes. Similar to low-income urban housing units in Boston, MA [41], pyrethroids and PBO were detected in higher concentrations and used more frequently in our study homes compared to other pesticides. This finding is consistent with the fact that pyrethroid insecticide formulations for residential applications have largely replaced OP pesticide residential formulations [42, 43]. Although over 19, 000 kgs of permethrin were applied in Monterey County in 2006 for agricultural purposes [44], we did not observe significant differences in permethrin concentrations (or loadings) between locations. Allethrin and cypermethrin were also widely detected in most homes. Our findings suggest that home use likely contributed to the presence of pyrethroid pesticides in house dust since pyrethroids were commonly used indoors and negligible to no agricultural applications took place at the county level (except for permethrin). It is also possible that structural pest control applications influenced indoor detection of certain pyrethroids in some homes. For example, it is estimated that ~80% of the non-agricultural cypermethrin use reported in Alameda County in 2006 was for structural pest control [38]. The presence of pyrethroids in house dust is also consistent with their physical and chemical properties, including high octanol:water partition coefficient values (log Kow > 4.0) and low vapor pressures. To our knowledge, only two studies [10, 13] have measured pyrethroid dust concentrations in farmworker homes. Similar to the present study, permethrins were the most frequently detected pyrethroids indoors. Median cis- and trans-permethrin concentrations in our farmworker homes were higher than those observed in a previous study [10].

The detection of chlorthal-dimethyl in all Salinas farmworker homes and none of the Oakland urban homes is consistent with other Salinas Valley studies showing an

TABLE 6 Dust Concentrations for Select Organophosphorous Pesticides and Pyrethroids from Select U.S. Farmworker Studies (ng/g)[a].

Author	Population	Location	Collection method	Sampling Dates	Pesticides	LOD (ng/g)[b]	DF%	n	Range (ng/g)	Median	Mean (SD)
Harnly et al. 2009[c]	**Farmworkers (CHAMACOS longitudinal birth cohort)**	**Salinas Valley, CA**	**HVS3**	**2000-2002**	**Organophosphates:**						
					Chlorpyrifos	2	91	177-197	2.9-7850	74	NR
					Diazinon	2	86		4.7-2870	26	
					Pyrethroids:						
					cis-Permethrin	5	98		16-168000	344	
					trans-Permethrin	5	98		146-265000	467	
					Others:						
					Chlorthal-dimethyl	2	98		2.3-271	22	
Bradman et al. 2006[d]	Farmworkers	Salinas Valley, CA	HVS3	June-September 2002	Organophosphates:						

TABLE 8 (Continued)

Author	Population	Location	Collection method	Sampling Dates	Pesticides	LOD (ng/g)[b]	DF%	n	Range (ng/g)	Median	Mean (SD)
Harnly et al. 2009[c]	**Farmworkers (CHAMACOS longitudinal birth cohort)**	**Salinas Valley, CA**	**HVS3**	**2000-2002**	**Organophosphates:**						
					Chlorpyrifos	2	95	20	<LOD-1200	49	NR
					Diazinon		100		4-810	21	
					Malathion		20		<LOD-480	NR	
					Pyrethroids:						
					cis-Permethrin		100		13-2900	150	
					trans-Permethrin		100		22-5800	230	
					Others:						
					Chlorthal-dimethy		100		6.5-110	31	

TABLE 6 *(Continued)*

Author	Population	Location	Collection method	Sampling Dates	Pesticides	LOD (ng/g)[b]	DF%	n	Range (ng/g)	Median	Mean (SD)
Harnly et al. 2009[c]	**Farmworkers (CHAMACOS longitudinal birth cohort)**	**Salinas Valley, CA**	**HVS3**	**2000-2002**	**Organophosphates:**						
Rothlein et al. 2006	Farmworkers	Hood River, OR	HVS3	Summer 1999	Chlorpyrifos	10	92	26	<LOD-1200	130	200(240)
					Diazinon	10	77		<LOD-720	310	310(230)
					Malathion	10	81		<LOD-1400	180	380(400)
Curl et al. 2002	Agricultural Workers	Yakima Valley, Washington State	Nilfisk vacuum cleaner	June-September 1999	Chlorpyrifos	150	26	156	<LOD-2560	50	NR
					Diazinon	170	3.8		<LOD-770	10	
					Malathion	160	15		<LOD-1030	40	

Author	Population	Location	Collection method	Sampling Dates	Pesticides	LOD (ng/g)[b]	DF%	n	Range (ng/g)	Median	Mean (SD)
Harnly et al. 2009[c]	**Farmworkers (CHAMACOS longitudinal birth cohort)**	**Salinas Valley, CA**	**HVS3**	**2000-2002**	**Organophosphates:**						
Fenske et al. 2002[e]	Ag (at least one family member employed as an orchard applicator (APP) or farmworker (FW))	Central Washington State (major tree fruit production region)	HVS3	May-July 1995	Chlorpyrifos	LOQ: 13-27(varied batch to batch)	APP: 100 FW: 100	APP: 49 FW: 12	APP: 10-2600 FW: 70-560	APP: 370 FW: 250	APP: 550(580) FW: 270(180)
Simcox et al. 1995	Farmers (F), Farmworkers (FW)[e]	Wenatchi area (eastern Washington State)	HVS3	Jan-July 1992	Chlorpyrifos	LOD: 20 ng/mL LOQ: 17 ng/g	F: 96 FW:100	F: 26 FW: 22	F: <LOD-3585 FW: 40-2180	F: 372 FW: 172	F: 506 FW: 338 SD not reported

a. Other studies may have measured additional analytes; only those relevant to the ones measured in our study are included. b. Unless otherwise indicated, limits of detection (LODs) are in ng/g. In some cases a limit of quantitation (LOQ) was reported instead of an LOD. c. Other analytes were also measured, but detection frequencies were <50%. Analytes included: malathion, methidathion, and iprodione. Minimum value reported in the "Range" column is the lowest quantified concentration. d. Other analytes were also measured, but detection frequencies were <50%. Analytes included allethrin, bifenthrin, cypermethrin, deltamethrin, esfenvalerate, sumithrin, and iprodione. Malathion had a detection frequency <50% in this study, but respective information is presented for comparison with other farmworker studies. e. Also included a reference population, but information is not provided in this table. SD = Standard deviation; NR: Value not reported; <LOD: Indicates that value reported was below the limit of detection or not detected.

association between agricultural use and house dust contamination [13] and a positive correlation between outdoor and indoor air concentrations [10]. This herbicide had relatively high agricultural use (~ 33, 970 kgs) in the Salinas Valley and is not found in home-use pesticides. Chlorthal-dimethyl also has a high log Kow value and low vapor pressure, and may be bound to particulate matter at room temperature.

Over 16, 000 kgs of malathion and iprodione were used in 2006 for agricultural applications; however, they were not commonly detected in farmworker homes from the city of Salinas. For some of these pesticides, e.g., iprodione, LODs were higher than for other analytes. Other factors including physico-chemical properties, e.g., high vapor pressure and low log Kow values (≤3), may have resulted in lower detection frequencies. These pesticides were also not frequently detected in dust samples from our previous study in the city of Salinas [10].

This study has several limitations. Location differences in pesticide dust levels have been reported previously when using loadings rather than concentrations [21]; however, our small sample size limits statistical power and may have prevented us from observing statistically significant differences between locations for concentrations and/or loadings. Additionally, although homes with insufficient sample mass were demographically similar to those with adequate sample mass, exclusion of these homes may have introduced some bias and prevented us from detecting a difference in pesticide concentrations and/or loadings between locations. We also focused on low-income homes and thus the results may not be generalizable to other populations. Although estimated intakes for select pesticides were below EPA RfDs (i.e., HQ <1.0), it should not be concluded that intakes below RfDs are "acceptable" or free of any health risks. For example, recent studies have identified mechanisms of OP pesticide toxicity that were not considered in defining current U.S. EPA RfDs (e.g., suppressed expression of serotonin transporter genes) [45]. Moreover, RfDs do not account for differences in vulnerability to pesticide toxicity due to genetic factors, such as paraoxonase (PON1) polymorphisms [46]. Additionally, our intake calculations for pesticides do not account for other exposure pathways (e.g., inhalation or diet); nor did we consider that some children could have pica or other behaviors that could increase or decrease intake. Although we surveyed participants on their usage of pesticides indoors, we were not always able to corroborate whether formulation ingredients were present at high concentrations as the pesticide containers were not always available to confirm the active ingredients. Lastly, children in the homes sampled are clearly exposed to multiple indoor contaminants and our hazard evaluation does not account for exposure to complex mixtures.

CONCLUSIONS

Studies of contaminants in low-income homes, including our study, have been limited in sample size and, often, selection of participants has not been random. In addition, collection methods, analytical techniques, analytes measured, and timing of data collection differ. To our knowledge, only one other study has assessed indoor dust concentrations of pyrethroids in low-income homes in an urban setting [41]. Nonetheless, the results from these studies indicate that low-income children are potentially exposed to a mixture of pesticides. Agricultural pesticide use may contribute to addition al exposures to some pesticides in rural areas; historical or current residential use is also likely to contribute to ongoing exposures. Although children's non-dietary intake did not exceed U.S. EPA RfDs for select pesticides, this does not ensure that children are free of any health risks as RfDs have their own limitations, and the children may be exposed indoors via other pathways. The frequent pesticide use reported among participating households in this and previous studies of low-income homes [18, 19, 41] and high detection of several home-use pesticides in house dust suggests there is a need to educate families on the potential health impacts of pesticide use and effective integrated pest management strategies to control pests and reduce exposures to household occupants [42]. Particular at-risk populations are those living in households with poorer housing quality, where there may be greater needs for pest control [18, 19].

Additional research is needed to quantify exposures and potential health effects from these compounds, particularly frequently used pesticides such as pyrethroids. Such research should consider the complex mixture of chemicals found in indoor environments, include both environmental and biomonitoring measurements to assess cumulative exposures, and consider exposures in homes of different socioeconomic status.

KEYWORDS

- **Chlorthal-dimethyl**
- **Environmental Protection Agency's**
- **Organophosphorous**
- **Pesticide use reporting**
- **Surrogate recovery standards**

REFERENCES

1. Bouchard MF, Bellinger DC, Wright RO, Weisskopf MG. Attention-deficit/hyperactivity disorder and urinary metabolites of organophosphate pesticides. Pediatrics 2010; 125:e1270–1277.
2. Engel SM, Berkowitz GS, Barr DB, Teitelbaum SL, Siskind J, Meisel SJ, Wetmur JG, Wolff MS. Prenatal organophosphate metabolite and organochlorine levels and performance on the Brazelton Neonatal Behavioral Assessment Scale in a multiethnic pregnancy cohort. American Journal of Epidemiology 2007; 165:1397–1404.
3. Eskenazi B, Marks A, Bradman A, Harley K, Barr D, Johnson C, Morga N, Jewell NP. Organophosphate pesticide exposure and neurodevelopment in young Mexican-American children. Environmental Health Perspectives 2007; 115:792–798.
4. Marks A, Harley K, Bradman A, Kogut K, Johnson C, Barr D, Calderon N, Eskenazi B. Organophosphate pesticide exposure and attention in young Mexican-American children: The CHAMACOS Study. Environmental Health Perspectives 2010; 118:1768–1774.
5. Rauh VA, Garfinkel R, Perera FP, Andrews HF, Hoepner L, Barr DB, Whitehead R, Tang D, Whyatt RW. Impact of prenatal chlorpyrifos exposure on neurodevelopment in the first 3 years of life among inner-city children. Pediatrics 2006; 118:e1845–1859.
6. Whyatt RM, Rauh V, Barr DB, Camann DE, Andrews HF, Garfinkel R, Hoepner LA, Diaz D, Dietrich J, Reyes A. et al. Prenatal insecticide exposures and birth weight and length among an urban minority cohort. Environmental Health Perspectives 2004; 112:1125–1132.

7. Eskenazi B, Harley K, Bradman A, Weltzien E, Jewell NP, Barr DB, Furlong CE, Holland NT. Association of in utero organophosphate pesticide exposure and fetal growth and length of gestation in an agricultural population. Environmental Health Perspectives 2004; 112:1116–1124.
8. Imamura L, Hasegawa H, Kurashina K, Matsuno T, Tsuda M. Neonatal exposure of newborn mice to pyrethroid (permethrin) represses activity-dependent c-fos mRNA expression in cerebellum. Archives of Toxicology 2002; 76:392–397.
9. Nasuti C, Gabbianelli R, Falcioni ML, Di Stefano A, Sozio P, Cantalamessa F. Dopaminergic system modulation, behavioral changes, and oxidative stress after neonatal administration of pyrethroids. Toxicology 2007; 229:194–205.
10. Bradman A, Whitaker D, Quiros L, Castorina R, Henn BC, Nishioka M, Morgan J, Barr DB, Harnly M, Brisbin JA. et al. Pesticides and their metabolites in the homes and urine of farmworker children living in the Salinas Valley, CA. Journal Expo Sci Environmental Epidemiology 2006; 17:331–349.
11. Butte W, Heinzow B. Pollutants in house dust as indicators of indoor contamination. Review of Environmental Contaminants Toxicology 2002; 175:1–46.
12. Colt JS, Lubin J, Camann D, Davis S, Cerhan J, Severson RK, Cozen W, Hartge P. Comparison of pesticide levels in carpet dust and self-reported pest treatment practices in four US sites. Journal Expo Anal Environmental Epidemiology 2004; 14:74–83.
13. Harnly M, Bradman A, Nishioka M, McKone T, Smith D, McLaughlin R, Baird-Kavannah G, Castorina R, Eskenazi B. Pesticides in dust from homes in an agricultural area. Environmental Science Technology 2009; 43:8767–8774.
14. McCauley LA, Lasarev MR, Higgins G, Rothlein J, Muniz J, Ebbert C, Phillips J. Work characteristics and pesticide exposures among migrant agricultural families: a community-based research approach. Environmental Health Perspectives 2001; 109:533–538.
15. Roberts JW, Wallace LA, Camann DE, Dickey P, Gilbert SG, Lewis RG, Takaro TK. Monitoring and reducing exposure of infants to pollutants in house dust. Review of Environmental Contaminants Toxicology 2009; 201:1–39.
16. Rudel RA, Camann DE, Spengler JD, Korn LR, Brody JG. Phthalates, alkylphenols, pesticides, polybrominated diphenyl ethers, and other endocrine-disrupting compounds in indoor air and dust. Environmental Science Technology 2003; 37:4543–4553.
17. Simcox NJ, Fenske RA, Wolz SA, Lee IC, Kalman DA. Pesticides in household dust and soil: exposure pathways for children of agricultural families. Environmental Health Perspectives 1995 103:1126–1134.
18. Bradman A, Chevrier J, Tager I, Lipsett M, Sedgwick J, Macher J, Vargas AB, Cabrera EB Camacho JM, Weldon R. et al. Association of housing disrepair indicators with cockroach and rodent infestations in a cohort of pregnant Latina women and their children. Environmental Health Perspectives 2005; 113:1795–1801.
19. Whyatt RM, Camann DE, Kinney PL, Reyes A, Ramirez J, Dietrich J, Diaz D, Holmes D, Perer FP. Residential pesticide use during pregnancy among a cohort of urban minority women. Environmental Health Perspectives 2002; 110:507–514.
20. Lu C, Fenske RA, Simcox NJ, Kalman D. Pesticide exposure of children in an agricultural community: evidence of household proximity to farmland and take home exposure pathways. Environmental Resource 2000; 84:290–302.
21. Egeghy P, Sheldon LS, Fortmann RC, Stout DM, Tulve NS, Cohel-Hubal E, Melnyk LJ, Morga MM, Jones PA, Whitaker DA, Important exposure factors for children: An analysis of laborator and observations on data characterizing cumulative exposure to pesticides. Research Triangl Park, NC: National Exposure Research Laboratory Office of Research and Development; 200 http://www.epa.gov/nerl/research/data/exposure-factors.pdf
22. Curl CL, Fenske RA, Kissel JC, Shirai JH, Moate TF, Griffith W, Coronado G, Thompson E Evaluation of take-home organophosphorus pesticide exposure among agricultural workers an their children. Environmental Health Perspectives 2002; 110:A787–792.

23. Gunier RB, Harnly ME, Reynolds P, Hertz A, Von Behren J. Agricultural pesticide use in California: pesticide prioritization, use densities, and population distributions for a childhood cancer study. Environmental Health Perspectives 2001; 109:1071–1078.
24. U.S. EPA. Chlorpyrifos revised risk assessment and agreement with registrants. Washington, DC: Office of Prevention, Pesticides, and Toxic Substances; 2000. http://www.epa.gov/pesticides/op/chlorpyrifos/agreement.pdf.
25. U.S. EPA. Diazinon revised risk assessment and agreement with registrants. Washington, DC: Office of Prevention, Pesticides, and Toxic Substances; 2001. http://www.ok.gov/~okag/forms/cps/epaagree.pdf.
26. Whyatt RM, Garfinkel R, Hoepner LA, Holmes D, Borjas M, Williams MK, Reyes A, Rauh V, Perera FP, Camann DE. Within- and between-home variability in indoor-air insecticide levels during pregnancy among an inner-city cohort from New York City. Environmental Health Perspectives 2007; 115:383–389.
27. Egeghy PP, Quackenboss JJ, Catlin S, Ryan PB. Determinants of temporal variability in NHEX-AS-Maryland environmental concentrations, exposures, and biomarkers. Journal of Expo Anal Environmental Epidemiology 2005; 15:388–397.
28. ASTM. ASTM-D-5438-94 Standard practice for collection of floor dust for chemical analysis. Annual Book of ASTM Standards, pp. 570–571. Philadephia, PA: American Society for Testing and Materials; 1994.
29. California Department of Pesticide Regulation. Pesticide Use Reporting Database. http://www.cdpr.ca.gov/docs/pur/purmain.htm.
30. Hornung RW, Reed LD. Estimation of average concentration in the presence of nondetectable values. Appl Occup Env Hyg 1990; 5:46–51.
31. Hedeker D, Gibbons RD. Longitudinal data analysis. Patterson, NJ: John Wiley & Sons; 2006.
32. Bearer CF. How are children different from adults? Environmental Health Perspectives. 1995; 103:7–12.
33. U.S. EPA. U.S. EPA. 2008 Child-specific exposure factors handbook (final report). Washington, DC: U.S. Environmental Protection Agency, EPA/600/R-06/096F. http://www.epa.gov/fedrgstr/EPA-IMPACT/2008/October/Day-30/i25908.htm.
34. FQPA 1996. Food Quality Protection Act of 1996. Public Law; 1996.
35. Barr DB. Biomonitoring of exposure to pesticides. Journal of Chemical Health and Safety 2008; 15:20–29.
36. Lewis RG, Fortmann RC, Camann DE. Evaluation of methods for monitoring the potential exposure of small children to pesticides in the residential environment. Archives of Environmental Contaminants Toxicology 1994; 26:37–46.
37. Reregistration Eligibility Decision for Phosmet (EPA 738-R-01-010) http://www.epa.gov/oppsrrd1/reregistration/REDs/phosmet_ired.pdf.
38. California Department of Pesticide Regulation. 2006 Annual Pesticide Use Report Indexed by Chemical-Alameda County. http://www.cdpr.ca.gov/docs/pur/pur06rep/chemcnty/alamed06_ai.pdf.
39. Fenske RA, Lu C, Barr D, Needham L. Children's exposure to chlorpyrifos and parathion in an agricultural community in central Washington State. Environmental Health Perspectives 2002; 110:549–553.
40. Rothlein J, Rohlman D, Lasarev M, Phillips J, Muniz J, McCauley L. Organophosphate pesticide exposure and neurobehavioral performance in agricultural and non-agricultural Hispanic workers. Environmental Health Perspectives 2006; 114:691–696.
41. Julien R, Adamkiewicz G, Levy JI, Bennett D, Nishioka M, and Spengler JD. Pesticide loadings of select organophosphate and pyrethroid pesticides in urban public housing. J Expo Sci Environ Epidemiol. 2008; 18:167–174.
42. Williams MK, Barr DB, Camann DE, Cruz LA, Carlton EJ, Borjas M, Reyes A, Evans D, Kinney PL, Whitehead RD Jr. et al. An intervention to reduce residential insecticide exposure during pregnancy among an inner-city cohort. Environmental Health Perspectives 2006; 114:1684–1689.

43. Bekarian N, Payne-Sturges D, Edmondson S, Chism B, Woodruff TJ. Use of point-of-sale data to track usage patterns of residential pesticides: methodology development. Environmental Health 2006; 5:15.
44. California Department of Pesticide Regulation. 2006 Annual Pesticide Use Report-Monterey County. http://www.cdpr.ca.gov/docs/pur/pur06rep/chemcnty/monter06_ai.pdf.
45. Slotkin TA, Seidler FJ. Developmental neurotoxicants target neurodifferentiation into the serotonin phenotype: Chlorpyrifos, diazinon, dieldrin and divalent nickel. Toxicol Appl Pharmacology 2008; 233:211–219.
46. Holland N, Furlong C, Bastaki M, Richter R, Bradman A, Huen K, Beckman K, Eskenazi B. Paraoxonase polymorphisms, haplotypes, and enzyme activity in Latino mothers and newborns. Environmental Health Perspectives 2006; 114:985–991.

Authors' Notes

CHAPTER 1

We thank R. Kaufman, D. Brody, and M.J. Brown of the U.S. Centers for Disease Control and Prevention for their work developing the lead-related NHANES survey questions.

The U.S. Department of the Housing and Urban Development funded this project under contract C-PHI-00931.

The opinions expressed in this paper do not necessarily represent those of the U.S. government.

CHAPTER 2

Acknowledgments
We thank Raisa Stolyar for support with chemical analyses. This investigation was supported by a pilot project grant from the Harvard National Institute of Environmental Health Sciences Center for Environmental Health, Grant number P30ES000002Authors' Contributions

Robert F. Herrick designed the study and was responsible its overall conduct; John D. Meeker was involved in the analysis and interpretation of the data and participated in the manuscript preparation; Larisa Altshal conducted the sample analysis and data compilation. All authors read and approved the final manuscript.

COMPETING INTERESTS

The authors declare that they have no competing interests.

CHAPTER 3

This work was supported by grants R01 ES009718 and R01 ES016099 from the National Institute of Environmental Health Sciences

CHAPTER 5

Competing Interests
The authors declare that they have no competing interests.

Authors' Contributions

DB took the lead on the manuscript. He co-wrote the background and wrote the sections on asthma, lung function and cancer and the conclusions. JLD wrote the section on air pollutants near roadways and contributed substantially to the background. CR wrote the section on cardiovascular health. All authors participated in editing and refining the manuscript and all read it multiple times, including the final version.

Acknowledgments

We thank Wig Zamore for useful insights into the topic. The Jonathan M. Tisch College of Citizenship and Public Service partially supported the effort of Doug Brugge and Christine Rioux. Figure 1 was reproduced with permission of the publisher.

CHAPTER 6

Acknowledgments

Funding for this research was provided by Trane Residential Systems, Inc., Tyler, TX and Environmental Health & Engineering, Inc., Needham, MA.

Authors' Contributions

TAM conceived of the study, and participated in its design and coordination and helped to draft the manuscript. TM carried out the modeling efforts. JA carried out the data analysis. DLM participated in the design of the study and drafted the manuscript. All authors read and approved the final manuscript.

CHAPTER 7

Competing Interests

The authors declare that they have no competing interests.

Authors' Contributions

MHK carried out the modeling analysis. JGA and DLM assisted in the design of the study and helped to draft the manuscript. MPF provided emissions information and helped to draft the manuscript. JJM participated in its design, provided emissions information and helped to draft the manuscript. TAM conceived of the study, and participated in its design and coordination and helped to draft the manuscript. All authors read and approved the final manuscript.

Acknowledgments

This research was funded by Kaz, Inc., Southborough, MA and Environmental Health and Engineering, Inc., Needham, MA.

CHAPTER 8

Competing Interests

The authors declare that they have no competing interests.

Authors' Contributions

AB led the design of the research, drafted the paper, carried out simulations, sampling and measurement. MQco–led the design of the research guided paper writing, facilitated workplace site visits and conducted paper revisions. MP and DM provided critical input on simulations design, VOC measurements methods, and paper revisions. All authors approved the final manuscript.

Acknowledgments

The authors acknowledge Brian LaBrecque at the Analytic and Organic Laboratory at Harvard School of Public Health for performing chemical analyses with the EPA–TO–17. We acknowledge Catherine Galligan at Sustainable Hospitals Program, Lowell Center for Sustainable Production, at University of Massachusetts Lowell for her assistance with the worksite access for the observational analyses. This investigation was funded by two grants from National Institute for Occupational Safety and Health (NIOSH) a) grant R01 0H03744 as part of the National Occupational research Agenda (NORA); and b) grant No. T42/CCT122961–02, via Harvard School of Public Health, Education Research Center Pilot Project Award. Its contents are exclusively the responsibility of the authors and do not necessarily represent the official views of NIOSH.

CHAPTER 9

We thank J. Rico, J. Frelich, L. Godfrey-Bailey, L. Pothier, A. Trisini, R. Dadd, M. Silva, and J. Reidy.

We acknowledge the National Institute of Environmental Health Sciences grant ES09718, ES00002. National Institutes of Health training grant T32 ES07069 supported R.M.A. The authors declare they have no competing financial interests.

CHAPTER 10

Competing Interests

The authors declare that they have no competing interests.

Authors' Contributions

LQA: conceived of the study; participated in the design, coordination, and implementation of all study field activities; conducted the statistical analysis; and drafted the manuscript; AB: conceived of the study; participated in the design, coordination, and implementation of all study field activities; and helped to draft the manuscript; MN: responsible for laboratory analysis of dust samples and quality assurance and control, and helped to draft the laboratory analysis section of the manuscript; MEH: provided assistance with previous CHAMACOS data used in the analysis of temporal trends of phased-out pesticides and helped to draft the manuscript; AH: contributed to the statistical phase and helped to draft the data analysis and results section of the manuscript; TEM: helped to draft the manuscript; JF: responsible for cleaning the data and providing feedback on the manuscript; BE: conceived of the study; participated in the design, coordination, and implementation of all study field activities; and helped to draft the manuscript. All authors read and approved the final manuscript.

Acknowledgments

Work was supported by EPA (RD 83171001, Science to Achieve Results-STAR-Graduate Fellowship Program F5D30812), NIEHS (PO1ES009605), UC MEXUS, and the UC Berkeley Center for Latino Policy Research. This work is solely the responsibility of the authors and does not necessarily represent the official views of the funders. We thank our staff and community partners including Dr. Pescetti and the staff from

Clinica de la Raza for helping with recruitment efforts, our study participants, and Dr. Rupali Das, Dr. Katharine Hammond, Marta Lutsky, Dr. Mark Nicas, and Dr. Rosana Weldon for editorial comments.

Permissions

Chapter 1: Lead Exposure of U.S. Children from Residential Dust was originally published as "Exposure of U.S. Children to Residential Dust Lead, 1999–2004: I. Housing and Demographic Factors" in *Environmental Health Perspectives* 2009; 117(3): 468–474. Reprinted with permission.

Chapter 2: Teachers Working in PCB-Contaminated Schools was originally published as "Serum PCB Levels and Congener Profiles Among Teachers in PCB-Containing Schools: A Pilot Study" in *Environmental Health* 2011; 10:56. Reprinted with permission.

Chapter 3: Flame-Retardants' Effect on Hormone Levels and Semen Quality was originally published as "House Dust Concentrations of Organophosphate Flame-Retardants in Relation to Hormone Levels and Semen Quality Parameters" in *Environmental Health Perspectives* 2010; 118(3): 318–323. Reprinted with permission.

Chapter 4: Bacterial and Fungal Microbial Biomarkers in House Dust was originally published as "Home Characteristics as Predictors of Bacterial and Fungal Microbial Biomarkers in House Dust" in *Environmental Health Perspectives* 2011; 119(2): 189–195. Reprinted with permission.

Chapter 5: Pollutants from Vehicle Exhaust Near Highways was originally published as "Near-Highway Pollutants in Motor Vehicle Exhaust: A Review of Epidemiologic Evidence of Cardiac and Pulmonary Health Risks" in *Environmental Health* 2007; 6:23. Reprinted with permission.

Chapter 6: Asthma Triggers in Indoor Air was originally published as "Control of Asthma Triggers in Indoor Air with Air Cleaners: A Modeling Analysis" in Environmental Health 2008; 7:43. Reprinted with permission.

Chapter 7: Home Humidification and Influenza Virus Survival was originally published as "Modeling the Airborne Survival of Influenza Virus in a Residential Setting: The Impacts of Home Humidification" in Environmental Health 2010; 9:55. Reprinted with permission.

Chapter 8: Airborne Exposure from Common Cleaning Tasks was originally published as "Quantitative Assessment of Airborne Exposures Generated During Common Cleaning Tasks: A Pilot Study" in Environmental Health 2010; 9:76. Reprinted with permission.

Chapter 9: Phthalate Monoesters from Personal Care Products was originally published as "Personal Care Product Use Predicts Urinary Concentrations of Some Phthalate Monoesters" in Environmental Health Perspectives 2005; 113(11): 1530–1535. Reprinted with permission.

Chapter 10: Pesticides in House Dust was originally published as "Pesticides in House Dust from Urban and Farmworker Households in California: An Observational Measurement Study" in Environmental Health 2011; 10:19. Reprinted with permission.

Index

B

E

F

I

Q

R

S

T

W

For Product Safety Concerns and Information please contact our EU representative GPSR@taylorandfrancis.com
Taylor & Francis Verlag GmbH, Kaufingerstraße 24, 80331 München, Germany

www.ingramcontent.com/pod-product-compliance
Ingram Content Group UK Ltd.
Pitfield, Milton Keynes, MK11 3LW, UK
UKHW021833260926
475702UK00049B/508

* 9 7 8 1 9 2 6 8 9 5 2 0 8 *